pear
shaped

marlies bongers and
corien van zweden

pear
shaped

the life story of
your uterus

With illustrations by Gijs Klunder
Translated by Alice Tetley-Paul

GREYSTONE BOOKS
Vancouver/Berkeley/London

First published in English by Greystone Books in 2025
Copyright © 2022 by Marlies Bongers and Corien van Zweden
Originally published in Dutch as *Biografie van de baarmoeder* by Uitgeverij
De Arbeiderspers in 2022
Illustrations copyright © 2022 by Gijs Klunder
English translation copyright © 2025 by Alice Tetley-Paul

25 26 27 28 29   5 4 3 2 1

Greystone Books Ltd.
greystonebooks.com

Cataloguing data available from Library and Archives Canada
ISBN 978-1-77840-160-2 (cloth)
ISBN 978-1-77840-161-9 (epub)

Editing by Esther Hendriks
Editing for the English edition by Jennifer Croll
Proofreading by Jennifer Stewart
Indexing by Stephen Ullstrom
Jacket and text design by Javana Boothe

Jacket illustration by Nazarkru

Printed and bound in Canada on FSC® certified paper at Friesens. The FSC® label
means that materials used for the product have been responsibly sourced.

Greystone Books thanks the Canada Council for the Arts, the British Columbia Arts
Council, the Province of British Columbia through the Book Publishing Tax Credit, and
the Government of Canada for supporting our publishing activities.

Greystone Books gratefully acknowledges the support of
the Dutch Foundation for Literature.

This book is intended for informational purposes and does not constitute health or
medical advice. The author and the publisher accept no liability for any damages
arising as a result of the direct or indirect application of any element of the contents
of this book.

Canada

MIX
Paper | Supporting
responsible forestry
FSC
www.fsc.org   **FSC® C016245**

BRITISH
COLUMBIA

BRITISH COLUMBIA
ARTS COUNCIL
An agency of the Province of British Columbia

Nederlands
letterenfonds
dutch foundation
for literature

Canada Council   Conseil des arts
for the Arts   du Canada

Greystone Books gratefully acknowledges the xʷməθkʷəy̓əm (Musqueam),
Sḵwx̱wú7mesh (Squamish), and səlilwətaɬ (Tsleil-Waututh) peoples on
whose land our Vancouver head office is located.

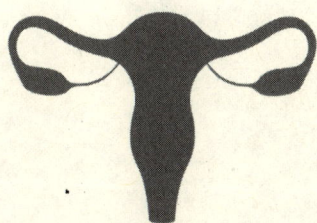

Contents

Foreword

THE UTERUS IS THE PLACE where we all began. It is, in essence, our first home. A dark and damp dwelling that gradually stretches as we develop from a bundle of cells into a fetus and all that goes with it. Until it can't stretch any further and the uterus decides it's time to give birth.

Childbirth is the uterus's main business and it's generally what comes to mind when we think of uteruses. That's hardly surprising, as the pregnant and birthing uterus puts on a spectacular performance: over the course of nine months, a hollow muscle not much bigger than the size of a pear becomes five hundred times its original size. In the process, it creates the housing in which an entire person can grow. And sometimes even two or more people at once.

Creating new life is an important and impressive task that the uterus traditionally carried out with gusto. But now, in the UK and Netherlands, women are giving birth to an average of just 1.49 children throughout their lives. While American women were giving birth to an average of 1.62 children in 2023, in 2022, the number in Canada dropped to the lowest level in history: 1.33 children per woman.

These numbers show the time the uterus spends on its core task has fallen significantly. On average, no more than 2 percent of a woman's life is devoted to being pregnant and giving birth.

For the remaining 98 percent of time, the uterus does all sorts of tasks, which vary depending on a person's phase of life. The uterus generally stays under the radar for the first twelve or thirteen years. It's busy growing in its hiding place in the pelvis. It first makes itself known during puberty, when it establishes the menstrual cycle in close cooperation with hormones. For most, it does this unrelentingly for the next forty years or so, unless any pregnancies occur. As menopause approaches, menstruation is terminated with the necessary fanfare, after which the uterus goes into retirement for the remaining thirty or so years of a person's life.

As long as a pregnancy doesn't occur, the uterus remains hidden between the bladder and the rectum. You can't see it or feel it from the outside. It usually only attracts attention because of its powerful ability to contract. If it does this as part of an orgasm, the contractions increase pleasure, although not everyone experiences this to the same intensity. The uterus's contractions during the menstrual period are more notorious. These can cause cramps that are reported as being uncomfortable or very uncomfortable by 60 to 80 percent of people who menstruate.

Uteruses are versatile organs. They can cause pleasure and pain, they can be a source of pride or shame, they can influence mood, well-being, and even identity. From a biological point of view, people assigned female at birth usually have uteruses, but not everyone with a uterus identifies as a woman. Transgender men and nonbinary people can also have a uterus.

Our aim with this book is to write the biography of an interesting and complex muscle that's so much more than just a birthing organ. We want to share the story of the uterus in all its different guises and roles. Its interesting anatomy that remained a mystery for centuries. The complex interplay of hormones. Varying degrees of bleeding. How fertility can be avoided if desired or stimulated if necessary. All the things that can go wrong with uteruses and with the complicated menstrual cycle we don't discuss enough. And all those years the uterus is not yet or no longer in operation.

We wrote this book first and foremost for anyone with a uterus, whether young or old, whatever gender they are, whether they are happy with the organ or even so much as detest it. But this book is, of course, also intended for anyone without a uterus who wants to know more about this ingenious organ and understand what having a uterus can entail.

By writing this book we hope to help eliminate taboos around discussing wombs, from menstrual bleeding, severe cramps, and conditions that affect uteruses to hot flashes and other symptoms of perimenopause.

Wouldn't it be great if girls didn't have to conceal their period pains or tampons? If a woman could tell her boss, without hesitation, that she'd be working from home because her period was heavy? And if another woman could ask, without embarrassment, for her working hours to be amended because she wasn't sleeping at night due to perimenopause?

Wouldn't it be great if we could talk about all things related to the uterus without shame and with full knowledge of the facts?

1

A Pear-Shaped Muscle in the Pelvic Cavity

On the anatomy of an ingenious organ and what can go wrong

IT TAKES A BIT of searching in a world of softly gleaming pinkish red, but then it's there, in full view: a tiny, perfectly round mouth that seems to be tightly closed. The camera carefully zooms in, and the mouth gets bigger and bigger until it fills almost the entire screen. Now a small opening is visible in the middle of the pink, fleshy rosette. The camera heads straight for it, works its way through the little gate. As the patient suppresses a cry—"Nice, deep breaths"—the camera enters a wide, orange-red space.

"This is the uterus."

Because the two of us are working on this book together, we spend a morning in Marlies's clinic in the Máxima Medical Center in the Netherlands. Marlies is standing in her white coat beside the patient's chair, calmly and routinely doing her work. Corien is leaning against the wall, ill at ease in a borrowed doctor's coat, with a notepad in her hand.

We focus on the screen. We watch as the camera probes the walls of the patient's uterus, circumvents a white floaty layer of mucous membrane and then stops at the opening of a fallopian tube. It's interesting to think just how new all of this is. We have a uterus in full view, sharp and clear at eight times magnification. The mysterious organ deep in the pelvis that remained hidden to all of our ancestors simply pops up on the screen in Dr. Marlies Bongers' clinic this September morning.

We look at it thoughtfully, the inside of that ingenious pear-shaped organ where we all started life. The uterus has been a source of intrigue for nearly all of human history, but no one had been able to see inside until recently. For centuries, very little was known about its exterior, its precise location, or its function. However, the uterus has long been the subject of theories and speculations, often put forward by men. Where was this organ that was usually hidden but showed itself when a child was on the way

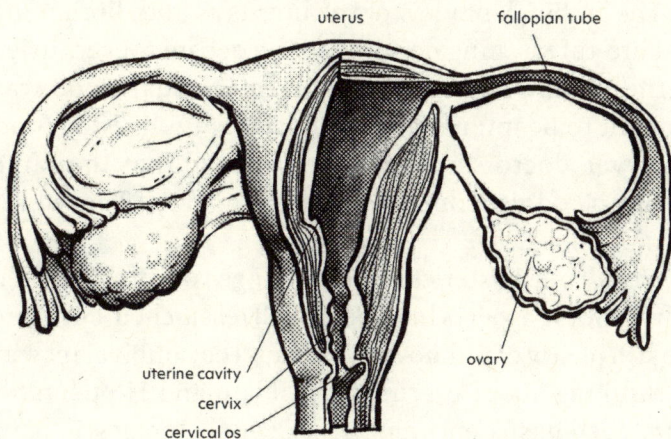

Uterus with two fallopian tubes and two ovaries.
The right side shows the inside of the uterus.

actually located? What exactly did it look like, what was its role, and why did it bleed every month?

A papyrus text dating from 1800 BCE contains the earliest reference to the uterus, calling it "the mother of mankind." Nevertheless, the ancient Egyptians described it as a somewhat complex and problematic organ. In the fourth century BCE, the philosopher Plato and the doctor Hippocrates further developed this line of thought. They described the uterus as ambivalent and at times dangerous. It was clear that the organ was involved in reproduction, but the role of menstruation proved a more complex question. In ancient times it was believed that monthly bleeding was associated with the removal of waste products and toxins. It was seen as a way of cleansing the female body: a process not required by the superior male body. Many of the ailments and illnesses that afflicted women were attributed to uteruses in which toxins and waste material were believed to have accumulated.

The ambivalent view of the uterus as both life-giving and life-threatening dominated the debate for centuries. In the Middle Ages, for example, menstrual blood was thought to be impure and to consist of undigested food. However, doctors at that time believed that the same impure blood also contained the pure material from which a child could grow.

Not all the mysteries surrounding the uterus have been solved. Some aspects have been well researched, but there are still plenty of unknowns. Perhaps that ambivalent way of thinking about uteruses that Plato and Hippocrates developed hasn't entirely left us. People today still have conflicting feelings and ideas about the uterus, which can switch between awe and embarrassment, and pride and shame.

Hysteria

In ancient Greece, the word for uterus was *hystera*, a term we still recognize in "hysteria" today. For centuries, all manner of inexplicable ailments that afflicted women— from pain to fits of rage, from depression to sexual disinhibition—were attributed to the uterus and therefore called hysteria. In the Middle Ages, hysteria was associated with wombs deprived of the offspring they desired. Women seen as hysterical were often also accused of witchcraft. These were mainly poorer, childless older women.

The psychiatrist Sigmund Freud breathed fresh life into the term hysteria by claiming that women who behaved hysterically—in whatever form that may have taken— were sexually unsatisfied. According to Freud, the illness was common and especially prevalent in widows, nuns, and unmarried women, but could be treated effectively with a so-called pelvic massage. Only male doctors were allowed to provide this type of treatment.

In the nineteenth century, hysteria became a fashionable disease. Those who could afford it would regularly see their doctor for a pelvic massage. As this type of a massage was quite an undertaking, doctors started experimenting with different tools. In 1880, the first vibrator became a reality.

View of the uterus

In ancient times, people had clear ideas about the anatomical shape of the uterus. As human autopsies were carried out very rarely, doctors mainly formed hypotheses based on feeling and looking, with the help of knowledge gained from dissecting dead mammals. According to Hippocrates, the uterus consisted of several chambers. This idea remained popular for many centuries. People had different opinions on the number of chambers, with some claiming it had as many as seven, but most believing it had no more than two. The Roman Greek doctor Galen, born in 129 BCE, claimed that the uterus had a left chamber for baby girls to grow in and a right chamber reserved for boys.

For many centuries, people didn't know how this mysterious organ with its various chambers was held in place inside the female body and where in the abdominal cavity it was located. Hippocrates developed the idea of the "wandering uterus." This was the belief that the organ could move freely throughout the abdominal cavity, "wandering" around the body. As a result, it could then encroach on other organs, leading to all sorts of issues. He believed this wandering was caused by an accumulation of semen and menstrual blood, which released vapors that could push the uterus up through the abdominal cavity.

By the Middle Ages, there was a simple and effective remedy for this affliction. In order to entice a wandering uterus back to where it belonged, scent therapy was often prescribed. If the uterus was thought to have moved upwards, foul-smelling burning wool mixed with sulfur, castoreum, and asphalt was placed below a woman's nose for her to sniff. This would repel the uterus and make it move back down again. Meanwhile, pleasant-smelling flowers would be placed near the vagina to entice the uterus back towards the pelvis.

Views on the uterus and its precise location intensified when people started carrying out autopsies of the human body in the sixteenth century. The first anatomical illustrations of uteruses, fallopian tubes, and ovaries began to appear in medical textbooks, becoming more precise over time. At first, these drawings were only intended for doctors and perhaps the occasional midwife, but that changed when printed material and books became more widely available and anyone could buy and read them. Nowadays all it takes is a few clicks of the mouse to conjure up not only detailed anatomical drawings, but also ultrasound images, 3D photos, or even videos of the uterus on our laptop screens.

Since ultrasounds, MRIs (magnetic resonance imagery), and laparoscopy became widely available in the 1980s and 1990s, women have been able to look at their *own* uteruses and ovaries if there are medical grounds to do so. And that's usually a different experience from studying an anatomical drawing of a uterus, however precise and complete that may be.

In her book *Are You on Your Period or Something?*, Dutch model and self-declared Menstruation Girl Lieke Smets recalls the first time she saw her uterus on an ultrasound scan. "All I could do was stare at the screen in awe. That was my uterus and those were my ovaries. Healthy and fertile. I was bursting with pride and couldn't stop smiling. 'If only I could take a photo,' I said. The gynecologist offered to print out the ultrasounds for me. I was so happy I could have cried."

Sanne (23) *on having two uteruses*

Sanne was only eighteen when she first went to the gynecologist with period problems; she was in severe pain and was bleeding continuously. Examinations revealed that she had a rare uterine abnormality: she'd been born with two uteruses. The left uterus had a regular cervix that appeared normal in the vagina. The body of the right one was normal but it wasn't clear whether it had a cervix.

As Sanne was still so young, laparoscopy and hysteroscopy were combined into one appointment and performed under general anesthesia. A laparoscope was inserted through the abdominal wall to see the outside of the uterus, while a hysteroscope was used to look inside the vagina. This gave the doctor a clear view of the patient's anatomy. It turned out there was a narrow opening to the right uterus in a fold of the vagina beside the cervix of the left uterus.

The doctor thought both uteruses might respond well to hormones to stop the heavy bleeding. Sanne tried four different contraceptive pills one after another, then the mini pill, then pills containing a higher dosage of progestins, and finally a vaginal ring (see chapter 3). Each time, things seemed to improve for a while, but the pain and bleeding returned a while later.

Sanne was at a loss. Nothing seemed to help, so Sanne and her doctor decided, as a last resort, to try injections to induce a temporary menopause. They hoped that this would finally put an end to the heavy bleeding. Sanne was enthusiastic at first, but she reacted badly to the injections. She no longer felt like herself and suffered from nightmares and hallucinations.

The situation seemed hopeless. What options were left? The gynecologist thought the right uterus might be the main culprit, as blood couldn't easily flow out of it, due to the abnormal shape and the way blood clots. To help the blood flow out, a very small, inverted funnel was inserted into the opening while Sanne was sedated. It seemed to work at first, but eventually the abdominal pains returned.

As the years passed, Sanne, who had completed her school exams and was now studying at college, still suffered from severe symptoms. The doctor suggested inserting a hormonal intrauterine device (IUD) in both uteruses. The side legs of the device intended for the right uterine cavity had to be clipped because the cavity was so narrow. However, this also proved unsuccessful.

For Sanne, one thing was clear after all this time: she couldn't carry on living with such severe symptoms. The impact on her life was too great. Something had to change. In consultation with her parents and a medical team, Sanne decided to have her right uterus removed. This could be done in such a way that the left uterus remained intact. The complex operation, which was performed with a laparoscope through the abdominal wall, was successful. Sanne's symptoms initially improved, but her problems didn't fully disappear.

Unfortunately, over time, the pain worsened again, and the heavy bleeding returned. After five years, Sanne asked for the other uterus to be removed as well. She explained that not a single day went by that she wasn't in severe pain. She wanted to lead a normal life, but her symptoms meant she often had to stay at home and miss out on parties, events, and trips.

Sanne's parents supported her decision. They had gone through everything with their daughter and knew how much the pain impacted her life. They were understanding and realistic and, like Sanne, couldn't see any other option. After numerous consultations and meetings with a psychologist, the decision was made to remove the left uterus too.

This operation was performed two years ago and every day Sanne is relieved. Her uterine abnormality was a terrible burden she was no longer able to bear. This young woman made an incredibly hard, but extremely courageous, decision.

Anatomy and shape

While there's no shortage of visual material of uteruses—you can even buy T-shirts with funny uterus images on them—by no means is everyone familiar with the structure of the organ. Gynecologists are used to describing it in simple terms. They may routinely draw sketches to help explain things or keep an anatomical model of the uterus on their desk. Sometimes that's a model of a nice, healthy, pear-shaped uterus with two elegant fallopian tubes attached. There are also models featuring all manner of tumors and polyps. Or it might even be a model with a fetus inside.

It's unsurprising that we're not particularly well acquainted with the organ, because from the outside—as long as a fetus isn't growing within—it remains invisible. A woman can only feel it when it cramps during menstruation or contracts during an orgasm. The uterus in its non-pregnant state is modestly sized and remains tucked away between the bladder and the rectum. In puberty, it grows from the size of a strawberry to that of a small pear. And it remains that size during the menstruating years as long as pregnancy doesn't occur.

The uterus is a hollow muscle with an incredible ability to stretch. It consists of smooth muscle tissue approximately two inches thick and can stretch to five hundred times its original volume during pregnancy. By the time it's at its largest towards the end of pregnancy, the muscle tissue has narrowed to just over an inch thick. After birth, the uterus immediately begins to shrink, and within a couple of weeks it has returned to the size of a pear. After menopause, when its working life is over, the uterus becomes smaller again until it's no larger than a strawberry.

The uterus has several parts. The largest—the round part of the pear—is the body of the uterus. The narrowest part is the cervix and measures an inch or inch and a half in length in non-pregnant women. The part where the cervix meets the vagina is called the cervical os (opening). The cervical os connects the uterus with the outside world via a small opening and sits in the vagina like a kind of spout. It is an extremely sensitive "mouth" full of many nerves that transmit pain, which, if triggered by too much pressure, can cause a woman to feel like she's about to faint.

An adult human's uterus weighs between one and four ounces and is held in place in the abdomen by bands of tissue called ligaments. These ligaments mean it can't wander around, contrary to Hippocrates's belief. There are two ligaments on the rounded upper side of the pear, which attach the uterus to each side of the pelvis. In between is a thin membrane that provides extra reinforcement. Farther down, where the uterus starts to narrow, another two ligaments attach to the rear of the pelvis. These are also connected by a membrane, so the uterus appears like it is suspended in a hammock in the lower abdomen. The pelvic floor muscle also helps keep the uterus in place.

The ligaments that support the uterus are a special kind of ligament. Organs like the stomach and intestines are also attached to the abdominal cavity by similar bands of tissue, but because the uterine ligaments have to be able to withstand the impressive growth of the organ during pregnancy, they are especially strong and elastic.

Although the ligaments are present in everyone with a uterus, the position of the uterus can vary. There are three possibilities: the uterus can be tilted forwards, it can sit right above the vagina in an upright position, or it can be tilted backwards. In the majority of white women, the uterus is tilted forwards. It was long thought that a uterus that was tilted backwards meant that a woman would have more trouble conceiving. At the start of the twentieth century, in Western Europe and America it was common practice to make women undergo an operation to tilt the uterus forwards. This only ended when it was discovered that the majority of Asian women had a uterus that was tilted backwards and that they didn't generally have any trouble getting pregnant.

Abnormalities

We now know that Hippocrates's belief that a uterus has several chambers isn't true. However, sometimes uteruses do develop with two chambers. In this case, there is a partition between the two parts of the uterus, resulting in a double uterus. This is caused by design errors that occurred in the early stages of embryonic development.

With female embryos, if all goes to plan, the uterus and fallopian tubes develop between weeks eight and thirteen. This involves two ducts fusing together to become one uterus, one vagina, and two fallopian tubes. Things can go wrong during this process. It may be the case that only a small partition is left, resulting in a minor abnormality (bicornuate uterus). But sometimes a larger partition results in two separate parts of the uterus. That's called a septate uterus. Sometimes each cavity has its own cervix with an opening to the vagina and both uteruses menstruate. If one of the two cavities doesn't have an exit, the menstrual blood has nowhere to go, which can lead to severe abdominal symptoms.

Other abnormalities can also arise. Sometimes the uterus and fallopian tubes fail to develop altogether during the embryonic phase, while the child-to-be has all the other female sexual characteristics, including a vagina. That's called Mayer-Rokitansky-Küster-Hauser (MRKH) syndrome and occurs in one in five thousand women. As the vagina develops normally and nothing appears abnormal externally, it's often only discovered when girls enter puberty, or sometimes even later. A girl born without a uterus will never menstruate, for example.

MRKH syndrome is a serious diagnosis, but it's not very common. If a girl's menstrual cycle doesn't start, that doesn't necessarily mean she doesn't have a uterus and fallopian tubes. There could be all sorts of other causes (see also chapter 2). A thick hymen without an opening could, for example, result in the absence of visible menstruation. The girl in question would be menstruating, but the blood would accumulate as it wouldn't have anywhere to go. This would lead to serious abdominal discomfort. It's simple to resolve this issue by carefully opening up the hymen.

It's relatively easy these days to detect major and minor abnormalities of the uterus with the help of imaging techniques like ultrasound, MRI, or hysteroscopy, but these methods haven't always been around. The seventeenth-century surgeon Job van Meekeren, for example, reported the case of a twenty-two-year-old woman who had never menstruated and suffered from unbearable abdominal pain. She was in so much pain "that she frantically tore her clothes to shreds as if possessed." Van Meekeren examined her and discovered that she had a very thick hymen without an opening. He used a sharp scalpel to open the hymen, after which "4.5 pounds of liquid the color and consistency of ground liver" left the body. The same night the girl lost even more old blood, but after that she was cured. She started menstruating normally and later became pregnant without any problems.

Wendy (51) *was born without a uterus*

As adolescents, Wendy and her friends eagerly awaited the arrival of their periods. One by one, her friends got theirs. When she was the only girl left in her class who hadn't got her period by the age of sixteen, she went to see her doctor.

Wendy was referred to the hospital, where they took blood and performed an ultrasound. A few days later, Wendy and her mother went to see the gynecologist for the results. "You don't have a uterus," he said bluntly.

"That's just the way you were born. There's nothing we can do." The doctor continued talking, explaining that she had ovaries and a very short vagina, but Wendy had stopped listening. Her ears were ringing and she felt queasy. All she wanted was to get away from there as quickly as possible.

When she got home, she burst into tears. She was devastated and spent a lot of time crying. It took months before she could bring herself to find out more about her diagnosis. She could feel for herself how short her vagina was. It made her anxious about getting a boyfriend. She was embarrassed about her abnormality and felt incomplete. She barely talked about it. And she ignored the advice to stretch her vagina by inserting smooth glass cylinders. It was too much.

It was only at the age of thirty-five that Wendy found a partner with whom she felt comfortable sharing her secret. Once she was with a man who loved her the way she was, she cautiously started experimenting with the glass cylinders. It went well. Ten years later Wendy discovered a swelling in her abdomen, which was causing problems when it came to urinating. She really had to force herself to go back to the gynecologist. Fortunately, this doctor was understanding and sympathetic. The gynecologist listened openly and let her tell her whole story.

The swelling turned out to be a fibroid. How was that possible? The gynecologist explained that she had a very small bud of the uterus, which had caused the growth of the fibroid. It all seemed so unfair to Wendy; the tiny piece of uterus she had was now making her life a misery.

Wendy's distress about having MRKH syndrome resurfaced with full intensity. The fibroid was removed during laparoscopic surgery. The intervention was successful, but it took a long time for Wendy to get her life back on track again.

Wendy is now fifty-one and is going through perimenopause. Of course, the hot flashes are a pain, but at the same time, she's pleased to be experiencing the same things as her friends with uteruses. "I finally feel like a real woman," she beams.

A view inside

The inside of the uterus is coated with a mucous membrane called the endometrium. It's a thin layer that measures around two millimeters ($^1/_{14}$ of an inch) thick in the years preceding the first period and in the phase of life after menopause, but it's often thicker during the fertile years, varying throughout the menstrual cycle. Just after menstruation, the endometrium is two or three millimeters thick, but it reaches a thickness of around six millimeters under the influence of estrogen around the time of ovulation. In the second half of the cycle, the endometrium has a better blood supply and reaches a thickness of six to twelve millimeters (close to half an inch). This forms a nice bed in which the fertilized egg cell can implant itself. If fertilization doesn't occur, the endometrium is broken down again and shed during menstruation.

The vagina also has a mucous membrane on the inside, but it looks different to that of the uterus. The vaginal mucous membrane is thin and smooth and resembles the mucous membrane in the oral cavity. It's pink in color.

The endometrium in the uterus is thicker and cloudier and consists of a large number of mucous glands. The color is somewhat redder than the vaginal mucous membrane. The two types of mucous membrane come together in the region of the cervical os, where the uterus meets the vagina.

Human papillomavirus (HPV) can infect the area where the layers of mucous membrane meet and overlap for a short distance. HPV is a virus that can be transferred via sexual contact. The body often deals with the virus itself by producing antibodies. However, in the event of long-term infection with HPV, changes in the mucous membrane around the cervix can occur, which can lead to cervical cancer.

Cervical screening (specifically, Pap tests) can detect these changes in the mucous membrane around the cervix at an early stage. In the examination, a small brush is used to take cells from inside the cervix exactly where the two layers of mucous membrane meet. These cells are examined in a laboratory for the presence of HPV and for abnormalities. If abnormalities in the cells are detected early, this precancer of the cervix can almost always be treated effectively.

Most Western countries have cervical screening programs, with screening recommended every three to five years. In the UK the program starts at age twenty-five, while in Canada and the US it begins earlier, at the age of twenty-one. The last screening in most places is somewhere between ages sixty-four and sixty-nine. Clinical testing is most common, although at-home tests are appearing in some countries. Since 2006, HPV vaccines have been available. In the UK, the US, and Canada HPV vaccines are recommended for children between nine and thirteen years old. It is optimal for people to get the vaccine before they become sexually active (though it is still possible to get the vaccine into adulthood). This vaccine gives 75 percent protection against cervical cancer. Once the vaccination rate is high enough, the frequency of the screening programs for cervical cancer may be decreased.

HPV vaccine programs are recommended for girls and boys, as boys can be infected with HPV as well. Of course, boys can't get cervical cancer, but HPV can also lead to throat or anal cancer. Furthermore, boys can spread HPV.

Fallopian tubes and ovaries

The uterus doesn't exist in isolation. It forms a unit together with the two fallopian tubes attached to it. In the sixteenth century, the Italian priest and anatomist Gabriele Fallopio (in Latin: Fallopius) was the first to describe the tiny tubes stretching from the ovaries to the uterus which now bear his name. The fallopian tubes are around four or five inches long and, like the uterus, are made of muscle tissue. However, their walls are much thinner than the walls of the uterus. The fallopian tubes are shaped like a funnel that can transport the eggs from the ovaries to the uterus.

There are still lots of unknowns about the fallopian tubes, which play a crucial role in fertilization. They are difficult to examine because they are thin and delicate; their internal diameter is no bigger than a third of an inch. The inner side of the fallopian tube has a thin mucus lining that is brought into motion by cilia. With the help of that mucus, the fallopian tube is able to "slurp up" the sperm cells that enter the uterus via the vagina, going against the flow. The passive flow through the fallopian tube goes the other way—from the ovary to the uterus—in order to transport a potentially fertilized egg towards the uterus.

The ovaries are two separate organs. Until well into the sixteenth century, ovaries were believed to contain female semen. Ovaries showed many similarities with testicles and were therefore simply called "testicles of women." At the start of the seventeenth century, the British doctor William Harvey was the first to claim that women didn't produce semen, but eggs. He also coined the phrase *omne vivum ex ovo*, or "all life comes from an egg." That's when the ovaries were given the name *ovaria*, from the Latin word *ovum*, or egg.

The ovaries are held in place in the abdominal cavity by their own ligaments. The ovaries—in older textbooks sometimes referred to as "egg nests"—are where egg cells are stored in follicles, small sacs of fluid, where they can grow and develop.

Although the ovaries are situated close to the funnels of the fallopian tubes, they can move independently of each other. Ovaries are immediately noticeable in the abdominal cavity; they are the only organs in the abdomen that are white in color. Ovaries change over a lifetime. In girls, they are tiny white dots or strands. In the fertile years, they develop into almond-shaped organs around an inch to inch and a half long. They have a bumpy appearance due to all of the follicles inside them. After menopause, when most of the hormonally responsive follicles have released eggs, the ovaries become much smaller. Though they do still function to produce testosterone, they are sometimes barely detectable by ultrasound.

In female embryos, ovaries are formed in week eight of the pregnancy. The ovaries initially contain millions of egg cells, but this number decreases over time. It takes a number of years before the first egg cell gets the chance to be released and to potentially be fertilized.

Various blood vessels supply blood to the uterus. The most important blood vessel is the uterine artery, which travels between the ligaments and branches off to form the many small blood vessels abundant in the uterus. This large, winding artery has a striking capacity for adaptation. It effortlessly provides a much higher blood supply to the pregnant uterus when it begins its period of spectacular growth. After the pregnancy has been completed at around nine months and the uterus begins to shrink again, the blood vessel adapts again.

This unique blood vessel that's absent from a male body is one of the reasons why it's not possible to transplant a uterus to a male body. You would have to transplant that special blood vessel along with it, which is no easy feat.

Alongside the uterine artery, two other arteries travel through the fallopian tubes towards the uterus and supply it with blood. As independent organs, the ovaries have their own blood vessel that travels from the side wall of the pelvis.

The uterus and sex

For people who have wombs, the organ plays a role in orgasms; it powerfully contracts a number of times and then fully relaxes. As orgasms can vary significantly between people, the role that the contractions play in orgasm will also differ. For many women and people with uteruses, however, the contractions of the uterus increase pleasure.

The contractions are not necessary to reach orgasm. Women who no longer have a uterus can also orgasm. People considering a hysterectomy may be told about the potential impact of the surgery on their sex life. If the operation is being carried out due to serious pain in the uterus, removal of the organ can perhaps even increase pleasure during sex. Sometimes no longer having to worry about getting pregnant can also improve pleasure during sex.

However, a hysterectomy can also potentially have a negative impact on a woman's sex life. Some women find it more difficult to orgasm without a uterus and others report a less intense orgasm. It's important for each person to reflect on these issues before a hysterectomy to make a well-informed decision.

There are various theories about why the uterus powerfully contracts during orgasm. Some people claim that the contractions create a propelling effect so that any potential sperm cells can be easily transported towards the fallopian tubes. Others claim that the total relaxation of the uterus after an orgasm facilitates the implantation of a fertilized egg. If that were the case, the relaxing effect on the uterus would have to be prolonged, as implantation generally only takes place a few days after fertilization.

Speculum

Various tools have been invented over the years to examine uteruses. One of the best-known instruments used during gynecological examinations is the speculum, which has a jaw that opens up like a duck's bill. Speculums date back to the ancient times, although the Roman speculums that were excavated in Pompeii looked more like glorified corkscrews with two spoons attached than what we are familiar with nowadays. Speculums today have two elongated arms and are often made of metal. As metal feels cold, the instrument is usually warmed with hot water prior to use.

Plastic, disposable speculums are also available. These feel more pleasant, but produce a great deal of waste. That's why an industrial designer from Delft in the Netherlands recently developed an environmentally friendly speculum made from sugarcane. This disposable speculum is either transparent or white and is available in four different sizes. Initial experiences have been very positive. It has the benefits of plastic but is completely biodegradable. Over the coming years we will have to carefully consider what the most sustainable option is: a modern, biodegradable speculum or the long-standing metal instrument that can be used time and time again.

The speculum allows the doctor to easily view and examine the vagina and the cervix. The instrument must be carefully inserted into the vagina, after which the two arms push the walls of the vagina apart so the doctor can look inside with the help of a light.

You hear all sorts of stories about speculums. Some women call them "medieval torture devices," while others have no issues experimenting with them at home. The famous book about women's health and sexuality, *Our Bodies, Ourselves*, which was first published in 1970, explained how you could examine yourself at home with the help of a plastic speculum, a flashlight, and a mirror. In that era of second-wave feminism, so-called self-help groups would get together to study their own and each other's vaginas.

To help doctors view the lower pelvis, a special chair was needed. A sixteenth-century German doctor was the first to design a type of gynecological examination chair. He lowered the back part of his operating table and installed foot supports. His idea was further developed over the following centuries. The gynecological examination chair with stirrups is the best-known—and perhaps also most notorious—design. In older chairs, still in use in some countries, the patient lies down on her back with her legs apart and bent, while her knees rest on the two supports. In newer chairs, patients can assume a more active position, sitting semi-upright with their feet on two supports.

Most gynecologists, even those who have been practicing their profession for many years and have carried out numerous gynecological examinations, are aware that patients often dread pelvic exams. They know that it will never be a particularly pleasant experience for most of their patients, so they do their best to make sure the patient is as comfortable as possible. In most cases patients are offered a gown or long shirt before the examination to cover their lower body.

Gynecological examinations can be somewhat uncomfortable, not only for patients but also for doctors. Young doctors are often a bit apprehensive about performing them. They usually learn the basic principles during their studies by practicing on plastic models of a pelvis. They sometimes also practice on volunteers. However, a doctor will usually remember the first time they carried out the examination for real as a nerve-wracking moment.

Gynecological examinations can be painful. Everyone who has had one knows that the cervix can be extremely sensitive to even the gentlest of touches. For younger women in particular, this can sometimes cause them to almost faint. This is a key feature of the stories people tell each other about having an IUD inserted. "It was a really horrific experience, but lots of people say that," twenty-four-year-old Jaimy recently recalled in an interview in a Dutch newspaper. The fainting response occurs because the body believes it may be in danger and sends all the blood to the muscles. As a result, the rest of the system doesn't get enough blood, which causes you to break out in a sweat and makes you more likely to faint.

Gynecologists are always mindful of that risk. That's why they ask patients not to get up too quickly after the procedure and remain lying down for a moment instead. Sometimes a patient may seem fine at first, but when they are standing at the desk booking their follow-up appointment, they suddenly feel awful. If the patient sits down for a while, puts their head between their legs and has a cup of coffee, they usually feel better again relatively quickly.

Uterus transplant

The first ever uterus transplant was the work of the Swedish gynecologist Mats Brännström, who moved to Australia for a few years in the 1990s to conduct post-doctoral research in gynecological cancer. There he treated twenty-seven-year-old Angela, who needed a hysterectomy due to a malignant tumor on the cervix. Angela asked her doctor, "Could I perhaps get a uterus transplant?"

That question shaped the career of Mats Brännström, who made uterus transplantation his life's work.

Worldwide around one in five thousand young women live without a uterus. Some suffer from MRKH syndrome, whereby a uterus fails to develop. Others have had their uterus removed early in life due to cancer or another condition.

Brännström ultimately managed to secure grants and get a team together to make uterus transplantation a reality. However, womb donation projects face many challenges. First of all, a good uterus donor has to be found. That's often the mother of the young woman without a uterus, or it might be a friend. If the mother wants to be a donor, the uterus in question usually belongs to a postmenopausal woman. In that case, the donor is given hormones in advance so that her small, shrunken uterus starts to grow again.

There are then two parts to the operation. First, the donor undergoes a hysterectomy, whereby the blood vessels attached to the uterus are preserved. This is a very challenging technique. During a second operation, the removed uterus is then transplanted into the recipient's body. The blood vessels need to be connected to the recipient's blood vessels in the pelvis. If it goes well, the transplanted uterus will menstruate for the first time around three months later. That's when it can be established whether or not the operation has been successful.

The question then is whether the transplanted uterus can carry a pregnancy. After a successful transplant, the likelihood of a pregnancy is around 80 percent. An embryo is developed in a test tube and then surgically placed in the transplanted uterus. In 2014, a child was born from a transplanted uterus for the first time. The baby was born a bit early via cesarean section as the doctors didn't want to wait until forty weeks. Since then, more than seventy children have been born from transplanted uteruses all over the world.

As with any transplant, the recipient needs to take strong medication after a uterus transplant to prevent rejection. Because of that, the transplanted uterus is often removed after one or two children have been born. The recipient can then stop taking the strong medication, which is much healthier in the long term.

Imaging techniques

It may have been possible in ancient times to examine the vagina and cervix with the help of a forerunner to the speculum, but what the uterus looked like from the inside remained a mystery for a long time. Various imaging techniques are now available to help us see within.

There's the ultrasound scan, which emits sound waves that are reflected back by organs. The images generated from these reflected waves are black, white, and various shades of gray. Ultrasound scans are commonly used during pregnancy to check the development and growth of the baby in the uterus. All that's needed to get a view of the baby in a more advanced pregnancy is to move the probe back and forth over the pregnant belly.

A non-pregnant uterus that's hidden away in the pelvis can be seen more easily with a transvaginal ultrasound where a narrow probe is inserted into the vagina. This technique is also effective in the early stages of a pregnancy, when the uterus is still small. To make the uterus clearly visible in women who aren't pregnant, a small amount of liquid can be infused into the uterine cavity through a small tube. This pushes the walls of the cavity apart. Liquid appears black on the ultrasound. If there is something like a polyp in the cavity, it will be clearly visible as the structure will be gray or nearly white.

Ultrasound images have improved in recent years due to technological developments, but patients don't always find the images easy to understand and interpret. You need a well-trained eye. With some help from a medical practitioner, however, a layperson can usually clearly see the pear-like shape of the uterus. The ovaries with their clusters of eggs can often also be made out.

In addition, MRI technology can be used. The MRI scanner uses magnetic fields and radio waves to display cross-section images of organs or tissues, slice by slice. A large uterus can be fully investigated with the help of an MRI. The MRI also clearly shows its position in relation to surrounding organs.

That morning when the pair of us—along with an assistant—are in Marlies's clinic, the patient's uterus is examined not only using a transvaginal ultrasound but also via hysteroscopy. This is carried out using a hysteroscope: an extremely long, thin tube that contains several very narrow channels. There's a channel for the camera and light and one for the saline that's needed to produce a good image. There's also another working channel that's big enough for tiny graspers or scissors to fit through.

The patient before us suffers from serious menstrual problems. Hopefully, the hysteroscopy will find an explanation. First, the hysteroscope needs to be carefully inserted into the uterus through the vagina and cervix. What many people don't know is that there are two "entrance gates" in the cervix that need to be passed: the external os and the internal os. The external os is the cervix's outermost opening and is more familiar to most, but the internal os—where the cervix meets the body of the uterus—is often the most problematic of the two.

Before the examination begins, the assistant offers some reassuring words. During the resulting moment of relaxation, the hysteroscope with the running warm saline is carefully but decisively inserted into the vagina. The external os is the first obstacle. Immediately after that comes the warning, "This may feel a bit uncomfortable. Take nice, deep breaths for me." That's when the hysteroscope passes through the internal os.

The moment of pain is quickly over, and the orangey red cavity of the uterus zooms into view.

2

The Uterus as Performer of the Monthly Cycle

On the complex interplay of hormones and the role of blood

ONE OF THE UTERUS'S JOBS is to carry out the menstrual cycle. It's an important, complex job that's been a source of much confusion throughout history. People were quick to understand that the uterus was involved in reproduction, but the purpose of the menstrual period proved a more complicated matter. All sorts of myths and tales about menstruation emerged. The menstrual period was already taboo in ancient times: on the one hand seen as dangerous, but at the same time mysterious or even magical.

In the first century CE, the Roman author and naval and army commander Pliny the Elder wrote, "Contact with [menstrual blood] turns new wine sour, crops touched by it become barren, grafts die, seeds in gardens are dried up, the fruits of trees fall off, the edge of steel and the gleam of ivory are dulled, hives of bees die, even bronze and iron are at once seized by rust, and a horrible smell fills the air. (...) There is no limit to the marvelous powers attributed to females. For, in the first place, hailstorms, they say, whirl-winds, and lightning even, will be scared away by a woman uncovering her body while her monthly courses are upon her."

To our modern ears, Pliny the Elder's words sound preposterous. We know better these days. But is that really the case? Today, despite all the medical knowledge that's available and despite a sexual revolution that took place some time ago now, menstruation is still often seen as taboo. Today, periods are still something to be ashamed of, something only spoken about euphemistically, if spoken about at all. The fables and myths surrounding menstruation have by no means disappeared.

Euphemistic language

It all starts with the euphemistic language so often used to talk about menstruation. Globally, there are around five thousand euphemisms, according to a questionnaire from the International Women's Health Coalition that was answered by 90,000 people in 190 different countries. The euphemisms range from relatively neutral expressions like "the time of the month" to more humorous terms or puns like "Aunt Flo," "Bloody Mary," "Bloody Sunday," "shark week," or "the curse."

Recent research shows that there are not only sensitivities around the word "menstruation" but also around the subject itself. A survey in the UK in *Happiful Magazine* reveals a gap in knowledge between men and women. One in ten British men have never had a conversation with a woman about periods. Period shame is a widespread phenomenon. Fifty-eight percent of young Canadian women find it shameful to talk about their periods and

40 percent of Dutch women report that they have made up an excuse to cancel an appointment because they didn't want to admit they were feeling unwell because of their period. And what are we to make of the fact that the word "menstruation" is often missing from the packaging of menstrual products and that ads invariably use blue—never red—liquid to show how absorptive the material is? Or the fact that Instagram bans a photo of a period stain on a pair of sweatpants? Or that girls complain they can't walk down their school corridors to the toilet with a tampon in their hand?

The taboo associated with menstruation is probably also one of the reasons why women often put up with serious menstrual problems for such a long time. They don't go to the doctor despite suffering from severe abdominal pain, heavy bleeding, serious headaches, or migraines. Although the symptoms occur in all age groups, they are generally most serious in women over the age of thirty-five. These symptoms aren't life-threatening, but they are common, usually occur every month, and can seriously affect quality of life.

Around 1.8 billion people worldwide have a period every month, which means that at any given moment in time eight hundred million people are menstruating. We don't know how many of them have menstrual pain, but statistics from the US give an indication. Forty-two million people in the US suffer from primary dysmenorrhea (menstrual cramps), and 3.5 million of them experience symptoms so severe that they are unable to work or go to school. This makes menstrual cramps the most common cause of lost school and working days among women in the United States.

You might wonder how different the situation would be if it were men, not women, who had to contend with periods and serious pain each month. Women don't generally call in sick, however much pain they're in. Each year, only one in twenty women go to the doctor with period problems. The remainder evidently keep quiet—perhaps they don't want to be seen as weak or perhaps they're embarrassed. Perhaps they don't want to complain, perhaps they're afraid to talk about it, perhaps they're ashamed, or perhaps their mothers told them it's just part and parcel of womanhood. They might not be aware that effective treatments are available for many of their problems.

The belief that pain and discomfort "are part and parcel of it" and just something you have to put up with as a woman is deeply entrenched and generally accepted to this day. In recent books—written by men—you still find the argument that women accept period pain as par for the course because they know that menstruation serves a higher purpose: producing offspring. Women who don't want children or are unable to have children should therefore simply accept the pain and discomfort of menstruation without complaint.

The first time

Let's start with the facts. What's the deal with uteruses and periods? The memory of the first time a girl gets her period will usually remain with her for the rest of her life. It's an important moment. Now that the two of us are having an in-depth conversation about menstruation in the context of this book, we recall our own experiences of "the first time." For Marlies it was a relatively relaxed moment—great, it has finally started—whereas Corien was in such severe pain that she had to stay home from school. Both of us had mothers who responded pragmatically by immediately supplying us with sanitary napkins and some explanation.

A person's very first period is called *menarche* and the body makes all sorts of preparations for it in the years preceding the big moment. When a girl approaches puberty, her body starts changing in various ways. Under the influence of the hormone estrogen, her uterus gets bigger, and the fallopian tubes, vagina, and labia also start to develop. These processes are hidden within the body and often go unnoticed. However, some changes are perceptible: breasts develop, hips widen, and pubic hair starts to grow.

Girls in the US and UK generally get their first period around age twelve. That's considerably earlier than it used to be. Around 1830, for example, the average age at which a girl got her first period was seventeen. Since then, this age has constantly fallen and that trend isn't showing any signs of stopping. While it would have been exceptional for a girl to start menstruating while in primary school in the 1950s, it would be nothing out of the ordinary today.

There are various theories as to why menarche is happening at an ever-younger age. The main explanation is better nutrition. If the body is well nourished, it is ready to become sexually mature earlier. Excess weight can further accelerate the process. Environmental pollution may also play a role. Endocrine disruptors that are present everywhere in our environment today—in drinking water, in packaging materials, in our food, in our car upholstery—may also be partially responsible. Furthermore, there seems to be a connection with smoking. The daughters of mothers who smoked during pregnancy tend to have an earlier menarche.

The first period marks the end of childhood and the start of the fertile years. This phase lasts until menopause signals the end of fertility around forty years later. During their fertile years, most people with a uterus menstruate more or less every month, unless they are pregnant or breastfeeding. That means that a woman has around four hundred menstrual cycles in her lifetime, on average.

The menstrual cycle—the word derives from the Latin word *mensis*, or month—lasts on average twenty-eight days. It's important to point out that that's an average, as cycle length can vary dramatically, not only between women, but also over the course of a lifetime and even from one month to the next. Some women have a cycle you could set your watch to, whereas others have an irregular pattern. Cycles that last between twenty-one and thirty-five days are all considered normal.

A cycle begins on the first day of the period and ends the day before the next period starts. Over the course of around twenty-eight days, egg cells in the ovary mature. One egg cell then becomes dominant and is released from the ovary. That's ovulation. The egg is usually collected by the fallopian tube's funnel. If fertilization doesn't occur in the subsequent twenty-four hours, the egg is removed as part of the menstrual period. From time to time, an egg that's just been released is not received by the fallopian tube. In that case, it ends up in the abdominal cavity and disappears.

The entire cycle is controlled by hormones produced in three different parts of the body: the hypothalamus in the midbrain, a gland at the base of the brain called the pituitary gland, and the two ovaries. There is a complex interplay of hormones whereby one hormone stimulates the other to do its work. These hormones not only influence the process of egg maturation, ovulation, and menstruation, but can also have an impact on a woman's well-being.

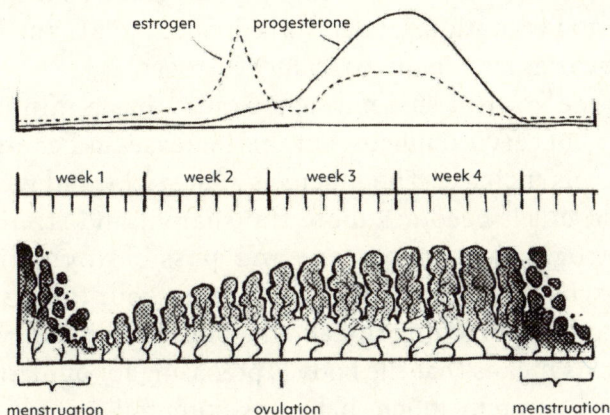

Schematic representation of the menstrual cycle.
The top figure shows the production of hormones during a cycle.
The bottom figure shows the thickness of the endometrium
in relation to the production of hormones

The cycle
Each cycle starts in the ovary, where egg cells have been biding their time since birth. These egg cells are miniscule and can only be seen under the microscope. Each egg cell consists of a nucleus in fluid, surrounded by a casing: the follicle. At the start of each cycle, a small group of egg cells in both ovaries prepares for ovulation—a competition that will ultimately only be won by a single egg cell, or two as

an exception. In the meantime, the quantity of fluid in the follicle increases.

This process from maturation to release of the egg cell is instigated by gonadotropin-releasing hormone (GnRH), a hormone in the hypothalamus that, in turn, encourages the pituitary gland to produce two hormones: FSH (follicle-stimulating hormone), which is responsible for egg maturation, and LH (luteinizing hormone), which sets ovulation in motion. Under the influence of FSH and LH, both ovaries also begin to produce estrogen.

At the start of this phase, a woman has a minimal amount of cervical mucus, but this increases as her estrogen levels increase. The mucus is a bit sticky and white at first, but it becomes more transparent and stretchy as estrogen levels increase and possible ovulation approaches. You can stretch it between your fingers to form a thin, glistening thread. The presence of this type of mucus means that the body is preparing for ovulation. The thin, stretchy mucus helps any potential sperm cells find their way inside.

By the time the process of egg maturation is complete, one egg cell in the ovary has become dominant and ready for ovulation. Under the influence of LH, the follicle bursts open and the mature egg cell is released from the ovary. The ovary then sends a signal to the pituitary gland to stop producing LH and FSH. At this point, estrogen levels peak. During this phase, women often feel energetic and enterprising. Ovulation marks the middle of the cycle. We have arrived at roughly day fourteen.

After ovulation, the mature egg is received by the funnel-shaped part of the fallopian tube, from where it can start its journey along the fallopian tube towards the uterus. The cilia on the inside of the fallopian tube help with this. From the moment of ovulation, the egg has around twenty-four hours to be fertilized. If fertilization occurs, this is usually during the first part of the journey through the fallopian tube.

Since the beginning of the cycle, the uterus has been busy preparing itself for the egg under the influence of estrogen. In order to ensure that a potential fertilized egg finds a good place to implant, the endometrium has been gradually thickening. Immediately after ovulation, when the ovary starts producing progesterone alongside estrogen, that process shifts into a higher gear. The endometrium becomes thicker and more compact. It even makes special adhesives that help the fertilized egg find a good spot. The progesterone also ensures that the opening to the cervix is blocked with a mucus plug. Vaginal discharge becomes stickier and drier. Everything is now in place for the fertilized egg to implant.

If it becomes clear at around day twenty-two that the egg has not been fertilized, the uterus starts to break down the thick, highly vascularized endometrium. In this phase, the ovaries produce more progesterone than estrogen. This causes libido to decrease, and the person may become more introspective or gloomy. The breasts may feel tender or swollen during this phase.

The day that the endometrium is shed is when menstruation begins. Estrogen and progesterone levels have become low, which can cause a woman to feel more emotional. The tension that may be present in the days prior to menstruation usually decreases as soon as the period starts. The uterus contracts to expel blood, which can cause pain that is sometimes severe. This pain is sometimes felt not only in the lower abdomen, but also in the back or thighs.

Menstruation usually lasts around five days. Each period, a woman loses between two to three tablespoons of blood on average. The menstrual fluid that consists of endometrium and blood also contains the non-fertilized egg. The blood can vary in color from bright red to dark red or brown, and the consistency can also vary. Sometimes it's a relatively thin liquid, but it can be thick and slimy, or it might contain red or liver-colored clots. That's all completely normal.

Over the course of a woman's fertile years, the quantity of blood loss usually increases. If a woman goes to the doctor with extreme blood loss, the first question is usually: How much blood are we talking about? Up to three and a half ounces is considered normal. A system has been devised whereby the number of used sanitary napkins or tampons per day is converted into a score that indicates whether a woman is losing too much blood. But as a doctor it's perhaps better to focus on what the woman is saying: if it seems like extreme blood loss, that must be the starting point.

Free bleeding

A small group of women who call themselves "free bleeders" completely reject the use of sanitary napkins and tampons. They consciously decide to simply let the blood flow. They have various reasons for doing this. For some, it's a feminist statement: they want to break taboos about menstruation and show that menstrual blood isn't dirty. Others believe that sanitary napkins and tampons contain toxins that are detrimental to our health. Some also see free bleeding as something spiritual: it helps them connect with their inner primitive self.

In Canada, Joelle Barron, a nonbinary writer, told the CBC that they have always preferred not to use a pad or tampon—"I just always was more comfortable without anything on." They explained that at home, they simply sit on a towel to manage flow. "I find my ability to grow people inside of me to be really, really cool," they continued. "Every time I get my period it feels like a celebration of that."

Collecting blood

There are many ways of collecting blood. Until disposable sanitary napkins were invented in the 1960s, women didn't have much choice other than to use cloths and rags. These had to be washed out and dried after each use so that they could be reused the next time. Special belts and straps were designed to hold the cloths in place. Later, so-called "sanitary bloomers" also came to the market, which were made of plastic and designed to prevent leaks.

The introduction of the first paper sanitary napkins represented a significant improvement for many women, although these pads were initially very thick and not especially absorbent. They were held in place with a belt or safety pins, as adhesive strips hadn't yet been invented. The "wings" that allow you to attach a pad to your underwear were only invented in the 1990s.

The great invention of the twentieth century was the tampon. The idea wasn't completely new; there's evidence that people had even experimented with inserting a rolled-up wad of material into the vagina in ancient Egypt. The first tampons as we know them today were developed in the years before the Second World War, but only came to the market after the war. The product wasn't immediately successful, but it steadily gained popularity from the 1970s onwards. In the United States, 70 percent of women now use tampons. It's important to replace tampons regularly—after a maximum of eight hours—due to the risk of toxic shock syndrome. This is a very rare but dangerous bacterial infection that can occur if a tampon is left in the body for too long.

The choice of menstrual products today is greater than ever; not only are tampons available in all sizes and materials, but sanitary napkins have continued to develop too; you can get ultra-thin but superabsorbent pads, mini pads, panty liners, and pads that are specially designed to be worn with thongs.

In addition, more sustainable alternatives are also being developed. The menstrual cup, for example, is gradually gaining popularity: a flexible, reusable cup that you insert into your vagina to collect blood and then empty out, wash, and reuse. An advantage of the menstrual cup is that you can easily see how much blood you are losing. One of the newest alternatives is the menstrual sponge. This is a small, reusable sponge that you insert like a tampon. Once inserted, it changes shape to fit your body, reducing the likelihood of leaks. Special period underwear is also available. It is made from an extremely absorbent material and can be reused after washing.

There's certainly plenty of choice these days when it comes to menstrual products, but these sophisticated items come with a price tag. In the US, women spend an average of twenty dollars a month on menstrual products—which adds up to eighteen thousand dollars over a lifetime.

This is a considerable sum, especially for families with multiple daughters and not a lot of money. Period poverty is a problem around the globe. In the UK, a survey carried out by the charity WaterAid revealed that a quarter of respondents struggled to afford menstrual products. As in the past, they are forced to walk around with cloths in their pants or to use wads of toilet paper. They often stay home from school or work due to shame and discomfort.

Various measures are being undertaken to combat period poverty. Many cities have begun offering free menstrual products—for example, in 2023, the Californian city of San Diego installed free menstrual hygiene dispensers in forty-eight locations throughout the city. Many food banks also stock menstrual products. However, our governments should really take care of this. Does it make sense, for example, that toilet paper is free in schools, but that menstrual products are not?

Fortunately, many high schools have now decided to supply free menstrual products in girls' restrooms. One school explained its decision to do so in a letter to parents: "Tampons and sanitary napkins are basic needs; it is essential that these needs are met in order to ensure adequate self-development. This means that pupils who experience period poverty will perform worse at school."

Seasons

Recently, things have started to change. Various parties are calling for the taboos around periods to finally be broken. It's generally young women who are fighting for this. As the author Nadya Okamoto says in her book, *Period Power: A Manifesto for the Menstrual Movement*, "Menstruation has been part of the human experience since the very beginning of our existence. Yet as a global community we have still not acknowledged access to menstrual hygiene as a universal natural need—or, TBH, acknowledged menstruation at all. It is a human right for every individual to feel equally capable and confident, respected, and even empowered, regardless of whether one is menstruating or not."

Another booked called *Period Power*, by the American author Maisie Hill, may have also inspired more acceptance of periods. This book, subtitled *Harness Your Hormones and Get Your Cycle Working for You*, was published in 2019 and became a hit. Hill's idea is to divide the menstrual cycle into four seasons: menstruation is winter, the phase leading up to ovulation is spring, ovulation itself takes place in summer, and, finally, fall is the premenstrual phase. Each season has its own characteristics, both physical as well as emotional. The idea is that if you take the seasons into consideration, your life becomes easier.

Groups of young women have adopted the new language usage and say things like "I'm not coming tonight, it's my autumn" or "I'm up for anything today, because it's almost my summer." Legions of women have started optimizing their lives around their cycles: as twenty-four-year-old Meg Diem told *The Wall Street Journal*, "Finding those patterns and using them to my advantage almost makes me feel like my period and cycle are superpowers that can enhance my life in different ways rather than defeat me."

Karin (45) *on severe premenstrual symptoms*

For Karin, the week before her period was always the worst week of her entire cycle. She thought it might be premenstrual syndrome (PMS). Recently, it had become almost unbearable. In the week before her period, she'd feel extremely angry and experience terrible thoughts. She didn't know what to do with herself, so made an appointment with her gynecologist. "You have to help me," she pleaded. "I get so aggressive just before my period. It's like a primal urge I can't suppress. I don't feel like myself at all."

The gynecologist asked how Karin was otherwise, whether there were any other issues and if she'd ever done anything about her menstrual symptoms before. Karin explained she'd been on the pill for a while, but didn't like it much. Because she was in a relationship with a woman, contraception wasn't necessary and she'd stopped taking it. She was an active woman and didn't have any serious issues in her life. She didn't have depression. As Karin's symptoms were so clearly linked to her cycle, the gynecologist diagnosed her with PMS.

One piece of advice Karin was given was to mark the week before her period in her calendar so that she could plan for it to be a bad week and to tell her partner too. During that week, Karin would have to take especially good care of herself; for example, she shouldn't schedule any important meetings and should eat well and get enough sleep. Of course, that didn't make her symptoms magically disappear, but it perhaps helped a bit.

In consultation with her doctor, Karin also decided to start taking a mild antidepressant drug during the second half of her cycle. Unfortunately, it didn't seem to work. Karin's feelings of anger and aggression kept returning each month despite the pills and advice.

In a follow-up consultation, the gynecologist suggested pausing Karin's cycle with the help of an injection. This would temporarily induce menopause, but Karin's cycle would return after three months. Karin went ahead with this, and after the tenth week, she visited the gynecologist again. She was so relieved. Her anger and aggression had simply vanished. However, she had started experiencing hot flashes and wasn't sleeping well. But this was nothing compared to the mood swings and terrible thoughts that had plagued her previously.

After much deliberation, Karin and her doctor decided to remove Karin's ovaries via laparoscopy to permanently induce menopause. It was a drastic measure, but after the success of the test procedure that induced an artificial menopause, it was clear how she would respond. After the procedure, Karin went to see the gynecologist one last time. She was delighted with her decision, even though she was now suffering from hot flashes. She was told she could take medication that would supplement small quantities of hormones—not in a cyclical manner like the pill, but the same dose each day. Karin decided she didn't want that and was happy the way things were. Despite the hot flashes, she felt a lot more balanced.

Premenstrual symptoms

Whether or not we divide the cycle into seasons, we think it's important that more attention is given to menstruation, the role of hormones in the function of the uterus, and the physical and mental symptoms that can arise from changing hormone levels. It remains to be seen whether hormones are always our best friends. Everyone who regularly menstruates or has menstruated at some point will be familiar with the negative feelings, grouchiness, and headaches that can occur in the days leading up to a period: the "fall." If these symptoms are so serious that they affect day-to-day functioning, it's referred to as premenstrual syndrome, or PMS. Though people may casually refer to PMS for even mild symptoms they experience before menstruation, it is a serious disorder which can really have an enormous effect on daily life.

PMS can occur if the body and mind react strongly to the coming period. PMS is most prevalent in women over the age of thirty-five whose menstrual cycles are already subtly beginning to change. These women are also more likely than younger women to suffer from heavy periods and spotting. PMS cannot be detected in the blood. Doctors rely on patients' accounts of the symptoms to make a diagnosis.

The fact that PMS cannot be physically detected can sometimes be problematic, for both doctors and patients. However, if you look at the pattern of symptoms, it's evident that PMS exists. Around 5 percent of women suffer from PMS, and that percentage is the same all over the world. Many of the feelings associated with PMS are similar to depression, but the difference is that the symptoms of PMS are clearly linked to the cycle; they appear in the second half of the cycle and recur every month. A very severe form of PMS is called premenstrual dysphoric disorder (PMDD).

Women are given all sorts of advice when it comes to dealing with PMS and PMDD: rest more, eat more healthily, do yoga, or meditate. If the symptoms are having a significant effect on day-to-day life, the gynecologist can also offer treatment options. The woman may decide to take medication that temporarily halts her cycle. If the symptoms disappear as a result, she'll know that it's PMS as opposed to depression or something else.

Halting the cycle and therefore inducing an artificial menopause is possible thanks to medication that works on the hypothalamus. The medication prevents the hypothalamus from causing the pituitary gland to produce FSH and LH, meaning the cycle doesn't start. One of the drawbacks of this treatment is that the woman may experience menopause symptoms, which is certainly not ideal. To prevent that from happening, low doses of estrogen and progesterone can be administered.

Another, more radical solution is to remove the ovaries. This will definitely resolve the symptoms of PMS, but the woman will instantly become menopausal. Furthermore, the procedure is irreversible. In the event of very serious symptoms, older women who do not (or no longer) wish to have children sometimes see this as the best option.

Absence of the cycle

There are other conditions related to the menstrual cycle beyond PMS and PMDD. A relatively common problem is the absence or cessation of the menstrual period, known as amenorrhea. It can be caused by various things, including the interplay of hormones or, as we saw in the previous chapter, an anatomical issue. For example, there may have been an error in the embryonic phase, which meant that the uterus or ovaries failed to develop.

More commonly, the cause of amenorrhea is related to body weight. For girls with a body weight that is too low—for example, due to an eating disorder—menstruation usually stops altogether. However, the cycle can also be disrupted by obesity. In addition, doing a sport to a high level may mean that the cycle doesn't start, or stops after a while. That's seen most often in endurance sports such as running, rowing, and cycling.

What women and girls don't generally realize is that the long-term absence of menstruation—which an athlete may even see as a positive thing—increases the risk of osteoporosis. If a girl or woman doesn't menstruate or no longer menstruates, her ovaries are not active, which means that her body produces much less estrogen. Estrogen prevents bone loss and progesterone is necessary to increase bone formation. Sports nutritionists should pay greater attention to this as people with eating disorders are also at risk of osteoporosis.

Stress can also play a role. High levels of stress can disrupt or stop the cycle, especially in young women. The absence of menstruation because of stress is sometimes seen in adolescent girls after their parents have gone through a difficult divorce, or after a close family member has died. Sometimes, menstruation can also be disrupted when a young woman first moves away from home for university. Often, such a disturbance only lasts a short time and the woman's periods usually return once the stress diminishes. It's not especially surprising that stress affects the menstrual cycle, as the part of the brain where stress is regulated is right next to the hypothalamus, which plays an important role in establishing the menstrual cycle.

Corien (59) *on PCOS*

My first period came late. All my friends already had theirs when mine finally arrived when I was sixteen. I'll always remember that first time as I was in so much pain that I had to go home from school. I remember signing myself out in the office. New flooring had just been installed. It stank so much I nearly threw up.

After that first time, my periods remained painful and very heavy. They were also completely irregular. My GP referred me to a gynecologist who prescribed relatively strong hormone pills to regulate my cycle. My cycle did become more regular, but the pills didn't help one bit with the pain. I stopped taking them a couple of years later.

I was twenty-five when I eventually went to see another gynecologist because I sometimes didn't have a period for months at a time and was still in a lot of pain. After a series of examinations, the doctor diagnosed me with polycystic ovary syndrome, or PCOS. I'd never heard of it and nobody around me knew anything about it either. The doctor explained that my irregular periods weren't problematic as long as I still had at least four or five periods a year.

Then the conversation took a turn. The doctor explained that various options were available to me if I wanted to have children. At that time, I hadn't considered whether or not I wanted children and hadn't even realized that the PCOS could affect fertility. I was mainly just relieved my symptoms didn't include obesity or facial hair, which my gynecologist had explained were common in women with this condition. I tried to live with the pain as best as I could.

When I went travelling around Southeast Asia for a few months at the age of twenty-nine, my abdominal pain got so bad that I arranged to see the gynecologist as soon as I returned home. We discovered I had a large cyst with a diameter of around three inches in one of my ovaries. It urgently needed to be removed. This was done in a clinic where in vitro fertilization (IVF) is carried out and where the walls are covered in birth announcements. I remember watching the cyst deflate like a little balloon on the ultrasound as it was punctured with a needle and drained. It was clear I'd been incredibly lucky. If the cyst had burst while I was away travelling, things could have turned out much worse.

In the years that followed, my menstrual cycle remained irregular. I had ultrasounds from time to time to check my ovaries for cysts. As the doctor only detected small cysts, no further treatment was necessary. I just had to make sure that the number of periods I had each year didn't drop too low, as then I'd have to start taking medication.

When I was thirty-three, my husband and I went to live in the tropics for a few years. We'd decided we'd like to start a family and had hoped to conceive. Unfortunately, my periods stopped altogether when we moved to northern Bolivia. After I'd gone ten months without a single period, we planned our first round of IVF to take place during our leave in the Netherlands. The day after landing at the airport, we had an appointment at the hospital. There, to our great delight, we found out that I'd become pregnant naturally!

After giving birth to a healthy daughter, I breastfed her for around fourteen months. It wasn't unusual that my period didn't come during that time. However, it was a huge surprise that the very first time I ovulated after stopping breastfeeding I became pregnant with a second daughter!

After two problem-free pregnancies, my cycle became regular for the first time in my life. Although my periods remained painful, the pattern characteristic of PCOS seemed to have disappeared. I feel incredibly lucky that I managed to become pregnant twice despite PCOS.

The absence of a regular menstrual cycle can also be indicative of PCOS, which is a hormonal imbalance. This condition affects 4 to 20 percent of women to some degree, making it the most common hormonal abnormality among women. The imbalance causes various follicles to start maturing at the same time in the first half of the cycle instead of a single egg cell becoming dominant and being released from the ovary. As ovulation doesn't take place, the rest of the cycle can't start and menstruation is absent. In the meantime, the follicles continue to develop and become sacs filled with liquid, or cysts.

Women with PCOS either do not have periods or have very irregular periods, and the various cysts in their ovaries can usually be seen clearly on an ultrasound. Sometimes there are as many as twelve follicles maturing in an ovary, while under usual circumstances there would be no more than eight. The various small cysts often look like a beaded necklace at the edge of the ovaries, but there may also be one or more very large cysts filled with liquid. If they are very large they may be painful. The cysts can be removed by piercing and then draining them, guided by ultrasound. Unfortunately, every now and then a cyst fills up with liquid again.

Ultrasound is not the only way to establish PCOS; it can also be detected in the blood. Women with this syndrome often have relatively high levels of the male hormone testosterone. PCOS often goes hand in hand with obesity; half of women with PCOS are overweight. They may also have hair growth on their face and chin. Furthermore, women may have reduced insulin sensitivity, which can ultimately lead to diabetes. Due to obesity and high testosterone levels, women with PCOS are also more likely to have high blood pressure, high cholesterol, and cardiovascular disorders.

Long-term absence of menstruation can increase the risk of uterine cancer. As ovulation doesn't occur, progesterone isn't produced, and the endometrium isn't shed. If this continues for many years, there's a risk that the cells in the endometrium may start growing abnormally, leading to cancer. This risk can be mitigated by medically inducing a menstrual period in women with PCOS a few times a year, which causes the endometrium to be shed. Fitting an IUD can ensure that the endometrium remains thin, meaning the growth cycle that can cause cell errors doesn't happen, making this a good option too.

Not much is known about the causes of PCOS. It sometimes runs in families, which suggests there may be a genetic link. As the number of people with obesity increases around the world, it's likely that the number of women who suffer from PCOS will also increase.

Christine (44) *on heavy menstrual bleeding*

Christine had always had heavy, painful periods. Starting from her first period at age thirteen, that was just the way it was, the same way it had been for her mom and aunties. She was used to taking painkillers, and it was completely normal for her to get up in the middle of the night to change her tampon or pad so she wouldn't end up with blood all over her bedding. "I'd tell myself: there's no use complaining about it. It's just part of who I am."

After the birth of her two children—Christine was thirty-four by then—her periods got even heavier. They were really painful and she would lose lots of blood that sometimes contained large clots. Christine also started feeling increasingly fatigued, but she brushed that aside. She had a baby that cried a lot at night; no wonder she felt tired. Her GP reassured her too; it was all completely normal.

The fatigue got worse. Christine looked pale, would get out of breath walking upstairs, suffered from hair loss and experienced leg and foot cramps at night. When her vision also started to deteriorate, she was referred to a neurologist. Fortunately, the brain scan revealed nothing of concern.

"It's my lack of fitness—I just need to do more exercise," Christine thought, pulling on her running gear despite her fatigue. It wasn't a success; she was so exhausted that she tripped and fell, and returned home with grazed knees. Even walking was starting to feel like hard work. She had to hold on to furniture as she walked around the room.

When she developed a high fever, she thought she might have contracted Lyme disease from a tick bite.

It was two days before her holiday to Sweden and she made an appointment with a doctor in the hope that she'd be able to get medication for it before going away. "We'll just check your iron too," the doctor said at the end of the consultation. That night, Christine's phone went off at half past nine. It was the doctor. "Sit down for a moment," he said. "You're so severely iron deficient that you need a blood transfusion immediately." Christine was shocked, but she protested; she really wanted to go on holiday. Could she not still go if she made sure to take things really easy?

The doctor agreed, but strongly advised her to start eating iron-rich food that night: lots of meat, preferably steak, and rye bread. As the shops were already closed, Christine's husband visited all the restaurants in the local area. At eleven o'clock at night, Christine forced down two plates of carpaccio and a steak. The next day she picked up some strong iron tablets from the pharmacy.

Christine spent most of her holiday in Sweden resting, while her husband took care of the baby and ran around after the toddler. With the help of iron tablets and an iron-rich diet, she slowly recovered over the course of the following months. To prevent the same situation from happening again, Christine had a hormonal IUD fitted. "It's much better now," she says. "The hormonal IUD means I don't get my period. It's not ideal, but I don't have any other choice."

Heavy menstrual bleeding

Heavy menstrual bleeding, or menorrhagia, is a common problem. Although 20 percent of American women contend with heavy menstrual bleeding, it doesn't get discussed much. Due to embarrassment, many women wait a long time before going to see a doctor about heavy periods, if they go at all.

In literature and films, you must look hard to find examples of women dealing with heavy periods. The French writer and philosopher Marie Cardinal formed an exception to this. Her autobiographical novel *The Words to Say It* (1975), in which she openly discussed her heavy blood loss, won a major literary prize and was turned into a film. In her book, she described how she bled almost continuously for years on end. "Until that strong and precise caress, very secret, very tender, would surprise me with a clot carried along by the blood. The dense, rushing lava, descending from a crater, invades the hollows, tumbling down, hot. (...) The blood had already had the time to reach my knees, even my feet, flowing from me in thin streams of beautiful, vivid red."

Cardinal's heavy bleeding, which appeared to have an emotional cause, led her to withdraw further and further from normal life. She became afraid to go anywhere. "How to avoid using the blood to explain that I could no longer live with others? I had stained so many easy chairs, straight-back chairs, sofas, couches, carpets, beds! I had left behind me so many puddles, spots, spotlets, splashes, and droplets, in so many living rooms, dining rooms, anterooms, halls, swimming pools, buses, and other places. I could no longer go out."

The cause of heavy menstrual bleeding is not always easy to establish. If someone has suffered from heavy periods since they were young, they may have a clotting disorder. Girls who have only just started menstruating sometimes suffer from menorrhagia around the menarche.

Furthermore, certain uterine abnormalities can also cause heavy bleeding.

In most other cases, heavy periods are probably linked to age. From the age of thirty-five, the cycle often starts changing before menopause. The hormonal balance gets disrupted, affecting how estrogen and progesterone are produced, which may result in a very thick endometrium. This can cause heavy periods, which may also last longer or be closer together. Research shows that 20 percent of women between the ages of thirty-five and fifty-five contend with heavy periods.

During a normal period, women lose a maximum of three and a half ounces of blood, but women with heavy periods often lose more than five ounces per period. Such heavy periods are not only inconvenient and sometimes plain disruptive—women are often forced to call in sick—but they can also lead to anemia. This is when the hemoglobin level in the blood becomes too low, which can lead to symptoms such as fatigue, shortness of breath on exertion, weakness, dizziness, headaches, palpitations, hair loss, and fainting.

These symptoms can gradually worsen over time. After a heavy period, the body only has three weeks to make up for the blood loss before the next period begins. Often, that's not long enough, so the hemoglobin level in the blood drops further with each subsequent period until the woman is seriously anemic.

Doctors aren't always quick to make the association with menstruation when women report symptoms such as fatigue, weakness, headaches, and dizziness. Often, women with these symptoms are referred to an internal health specialist; no one considers that anemia resulting from heavy periods might be causing these symptoms and that a referral to the gynecologist might make more sense.

Heavy menstrual flow can interfere with day-to-day functioning. Various treatment options are available to those who want them. In the first instance, a doctor may suggest the contraceptive pill or an IUD. Taking the pill means ovulation won't occur. As a result, the uterus won't receive the signal to build up a thick endometrium and heavy periods don't happen. The estrogen and progesterone in the pill keep the membrane thin. Having a hormonal IUD fitted that releases progesterone won't prevent ovulation, but it does mean that the uterus is unable to produce a thick endometrium. As a result, menstruation is sometimes almost completely absent.

For women with very heavy menstrual flow who don't wish to have children, endometrial ablation can be a good and often permanent solution. In this process, the endometrial lining of the uterus is removed in the doctor's office or in an operating room. This takes place under local anesthetic or light sedation. There are various methods of endometrial ablation. Often, a small fan-shaped device is inserted and opened inside the uterus. Once the device is in place, an electrical current flows across its surface, heating the endometrium so its cells are destroyed, right down to the deepest layer. The equipment stops automatically once the entire endometrial layer has been treated, ensuring that the uterus's muscular wall isn't damaged. In the past, a balloon containing hot liquid was sometimes used, which had the same effect on the endometrial cells.

After treatment, menstruation either reduces or completely stops. The treatment is often sufficient, but it can

be the case—especially with relatively young women—
that the endometrium starts to grow back again after a
while. It's not usually possible to perform a second endometrial ablation.

The most radical solution for heavy menstrual flow is
a hysterectomy. This is usually only suggested if all less-
invasive treatment methods have been attempted and
they failed to provide the desired result. But a hysterectomy may also be a woman's first choice. Sometimes a
woman comes into the office and says, "I've had enough. I
never want a period again. Please just take out my uterus."

Women differ greatly in this regard. Some women
are happy trying almost any treatment option, but a hysterectomy is out of the question. However serious their
symptoms, they don't want to lose their uterus. It's part
of them and it's inextricably linked to their identity as a
woman.

COVID-19 and menstruation

It's been a few years since COVID-19 gripped the world,
but people are still talking about how it affected periods.
Many people fell ill, and large numbers of people were vaccinated, both of which appeared to affect the menstrual
cycle. What's going on here?

Being sick with coronavirus only appears to have a
minor effect on menstruation. Of the women who were
in hospital—and therefore seriously ill—with COVID-
19, 25 percent reported menstrual changes. Most of them
reported lighter periods. After two months, their periods
were back to normal again.

A study carried out in the United States looked at the
effect of COVID-19 vaccination on menstruation. The call

to take part in the study was made via social media, which may have meant that most responses came from women who had noticed changes. Of respondents, 40 percent reported that their menstrual cycle had changed since vaccination, with heavier menstrual flow being the main symptom. At the end of December 2021, a foundation in the Netherlands tracking adverse drug reactions reported that they had received 17,000 reports of a change to menstrual flow. Most women had heavier periods after the vaccination.

However, menstrual changes because of vaccination appear to only be short term and menstruation usually returns to normal after a couple of months. The changes probably occur because the vaccination activates the immune system, in a similar way to what happens in the event of an infection. In the second half of the cycle, the immune system is usually slightly less active. If the immune system has been stimulated by vaccination, it can cause a change to the subtle process of building up and shedding the endometrium, which can result in heavier menstrual flow.

It's important to note that no study has yet suggested that the COVID-19 vaccination influences fertility. For the time being, we can conclude that some women experience a temporary change to their menstrual cycle after vaccination. This is something that also commonly occurs in times of tension and stress. It sorts itself out again.

Although the menstrual cycle can bring with it all sorts of problems, periods are an essential part of life for a person with a uterus. Menstruation also enables fertility. The menstrual cycle provides a rhythm, variable moods, peaks and troughs, pain, blood, and discomfort, but sometimes also strength and pleasure. Talking about it helps, as does taking it into consideration.

The latter is often hard enough. Can you call in sick with period problems or should you take a day off if you really can't work? Spain is the first European country to pass a law offering people with painful periods paid menstrual leave of up to three days. The Spanish Equality Minister, Irene Montero, called it a "historic day for advancing feminist rights."

It remains to be seen whether leave regulations for menstruation and menopause will come into force all over the world. But irrespective of this discussion, it's important not to hide or cover up the pain and problems associated with menstruation. Women often lack the knowledge to know which symptoms are normal and which aren't. Excessive pain and extreme blood loss are common, even if they're not talked about much. It's worth going to see a doctor if you have these symptoms, as a range of solutions and treatment options are available.

3

Preventing the Uterus From Doing Its Job

*On ways of avoiding, preventing,
or terminating pregnancy*

THE DESIRE TO UNDERSTAND the uterus and stimulate fertility is just as old as the opposite desire: to prevent conception. Or, as the American journalist Jonathan Eig put it in his book about the contraceptive pill: "For as long as men and women have been making babies, they've been trying not to." Contraception appears to be as old as humanity itself.

There are Egyptian medical papyri dating from 1850 BCE that discuss how to prevent pregnancy. It's doubtful the mixtures containing crocodile dung or vaginal rinses with vinegar the ancient Egyptians recommended were particularly effective. The many methods invented in the following centuries—ointments, drinks, objects in the vagina, linen sheaths to be worn on the penis—also proved rather ineffective. For centuries, unwanted pregnancies were the order of the day.

It was only in the late 1950s that a revolutionary discovery was made that would change the contraceptive scene for good. The invention of the contraceptive pill meant we could prevent ovulation and make pregnancy impossible by influencing the hormonal cycle with a daily pill. For the first time in history, women were given the power to control their own fertility. Conception could be prevented at will or postponed until further notice.

The special pills that were developed in the United States initially came to the market as medication to regulate the menstrual cycle, with temporary infertility as a side effect. It seemed safer to avoid the word "contraception" where possible, as birth control in the late 1950s was still a sensitive subject. In many US states and in various other countries, attempting to prevent a pregnancy was a criminal offence. The Roman Catholic Church, which rejected all forms of birth control, played a part in this.

In 1960, official permission was granted—despite protests from conservative America—to position the pills as a contraceptive, and things progressed rapidly from there. Five years later, six and a half million American women were taking the contraceptive pill that became known simply as "the pill."

The fact that women in many Western countries turned to the pill en masse from the 1960s onwards had unexpected consequences; the sexual revolution began and sex before marriage became increasingly common. The average number of children that a woman gave birth to in her lifetime dropped significantly. Now that they were in charge of their own fertility, many women opted to study or work instead of having lots of children. Many people saw the pill as liberating; they weren't aware that the long phase of development and testing prior to the introduction of the wonder drug had been linked to the major oppression of women.

Recent publications reveal that the contraceptive pill was tested on women under false pretenses and sometimes even under duress during its development phase. In the United States, for example, psychiatric patients were used as test subjects. In Puerto Rico, experiments were carried out on students and residents of shanty towns. The students were told that it would negatively affect their grades if they didn't take part in the study. It was easy to convince the mothers in Puerto Rico's shanty towns, in desperation due to their poverty and their numerous children, to take part in the tests, especially as any risks were kept vague.

Roos (24) *on making the difficult decision to use contraception*

Roos had period problems from a young age. Her periods were often very heavy and sometimes so painful she would faint. When she visited the doctor at age seventeen, she was prescribed the contraceptive pill for her symptoms. She didn't hesitate to take it. It was a practical decision more than anything. She had a boyfriend and was about to take her final exams. She didn't want her abdominal pain to get in the way.

The pill seemed to alleviate the pain for the first few months, but then it returned. It varied; sometimes her periods were manageable, but other times she needed lots of painkillers. She occasionally felt really emotional, but it was unclear whether that was because of the pill.

After two years, Roos stopped taking the pill. She was no longer in a relationship and wondered why she should take something every day that she didn't really need. The pill was having no effect on the pain anyway, which was the reason she'd started taking it in the first place.

Roos now knows much more about contraception than she did when she started taking the pill. For Roos, contraception is an interesting and important topic that comes up a lot in her studies and work. She often discusses the advantages and disadvantages of the pill and other forms of contraception with her friends. Roos believes it's important to look at it from all sides. She wonders whether all the negative opinions she hears about the pill are justified. "It's completely different for everyone," she says. "A lot of people want a natural cycle, but what if you have an unwanted pregnancy? The pill is still the best option in terms of reliability."

Due to persistent abdominal pain during her period, Roos has had an IUD for the last two years. Since then, her period pain has improved, but she still needs painkillers from time to time. She's surprised that contraception still seems to be a women's issue, but she also sees why nothing ever became of the male pill. It would only work if you could trust each other one hundred percent. Often, that's not the case.

Roos thinks there should be better sex education in high schools so that boys also start thinking about the subject. She also believes doctors should stop blindly prescribing the pill to girls without much consultation. "It's a complex subject and the choice you make is very personal. You need to be well informed and have a proper discussion about it first."

The pill

One of the people responsible for the development of the contraceptive pill was the American researcher Gregory Pincus. When Harvard University denied him tenure and refused his reappointment after the science fiction-wary public reacted badly to his successes with in vitro fertilization of rabbits, he founded his own scientific institute to pursue his research into hormones. Pincus was interested in the hormone progesterone, which is responsible for preventing ovulation during a pregnancy. He thought that administering a synthetic form of progesterone—progestin—to women in the form of a daily pill could perhaps cause temporary infertility. Surely the presence of progestin would cause the body to think it was pregnant, which would then prevent ovulation.

He was right, and Pincus's idea provided the basis for the first contraceptive pill. Those first pills contained relatively high doses of both female hormones: progestin and estrogen. As the pills turned out to have quite a few side effects, such as nausea, water retention, headaches, and migraines, a lower-dose pill was introduced in the 1970s. Developments in the field of contraceptive pills are still ongoing; we're now on the fourth generation of pills.

As most contraceptive pills contain both estrogen and progestin, they are referred to as the combined pill. The estrogen prevents breakthrough bleeding and supports the work of the progestin. The progestin, in turn, not only prevents ovulation but also ensures that the endometrium remains thin and that the cervix is closed off with a mucus plug. As ovulation doesn't take place, the ovary partially goes into an idle state, although everything remains in place so that another cycle can happen if the woman stops taking the pill. When taking the pill, egg cells still mature but ovulation doesn't take place.

The pill needs to be taken at around the same time each day, starting on the first day of menstruation. From day twenty-one, there's a seven-day break, in which the woman has a menstrual period. It's not a real period, as it's artificially induced. This withdrawal bleeding is usually lighter than a normal period.

A seven-day break is not strictly necessary. Women can choose to keep taking the pill, and in that case, they won't have a withdrawal bleed. Some people think this is unhealthy, assuming the uterus needs to get rid of the blood. However, this isn't the case. If a woman takes the pill, only a small amount of endometrium is produced on the inside of her uterus. Without a seven-day break, the membrane remains thin. Sometimes it's so thin that the blood vessels inside are vulnerable to damage, and it bleeds a little bit. That's breakthrough bleeding, or spotting. If breakthrough bleeding happens when you're taking the pill, you can decide either to have a seven-day break or to continue taking the pill and the bleeding will often stop on its own.

In the 1990s, a new contraceptive pill was developed that only contains progestin (i.e., it doesn't contain estrogen), known as the mini pill. This type of low-dose pill is popular among women who want to take as few hormones as possible. Women often take the mini pill after giving birth, as its effect on breast milk production is minimal.

The disadvantage of the mini pill is that it suppresses ovulation, but not very strongly. Women who take the mini pill must therefore be very precise when it comes to timing for it to be effective. Spotting is also more common with progestin-only pills.

However low-dose the pills may be, side effects are possible with every contraceptive pill. The list of side effects is long and includes headaches, migraines, breakthrough bleeding, breast pain, nausea, and weight gain. The pill is also linked to symptoms such as low libido and low mood or feelings of depression, and increases a person's risk of blood clots (thrombosis). Whether or not there are side effects varies from one woman to the next. It's often a case of trial and error. A pill that has hardly any side effects for one woman might lead to all sorts of problems for another.

Due to the increased risk of thrombosis, the use of the pill is often not recommended for women who smoke or have high blood pressure. Women who have had deep vein thrombosis are often advised not to take the pill unless another cause has been confirmed and something has been done about it. In those cases, the pill associated with the lowest risk of thrombosis is usually prescribed: the mini pill (progestin only). If the woman doesn't get on well with this pill, she could then try third-generation pills. There's plenty of choice. Most contain the same synthetic estrogen, but a different form of progestin.

Finally, the pill is not recommended for women who have or have had breast, ovarian, or uterine cancer. The hormones in the pill adversely impact hormone-sensitive tumors.

Other hormonal contraceptives

Other hormonal contraceptives have been developed alongside the oral contraceptive pill. For women who aren't keen on taking a tablet every day or worry that they might forget to take it, the contraceptive injection (Depo-Provera is a well-known brand) is an option. A GP can administer this injection once every three months. It works by gradually releasing a relatively high dose of progestin, which prevents ovulation. After halting the contraceptive injection, it can take a while for the menstrual cycle to return to normal.

Alternatively, a doctor or nurse can place a contraceptive implant known as Nexplanon under the skin in the upper arm. This plastic rod gradually releases progestin into the bloodstream, preventing ovulation for three years. As a result of the implant, menstruation can change and sometimes stop altogether.

Hormonal contraceptives are also available that release hormones on a localized level. The hormonal IUD (including brand names like Mirena or Kyleena) is the best known. It's a T-shaped plastic frame measuring just over an inch long that gradually releases progestin into the uterus. A doctor inserts it through the cervix into the uterus, where it can stay for around six years. The IUD thins the endometrium so that it becomes very difficult for a fertilized egg to implant. Women with an IUD usually have very light periods and 20 percent of them don't menstruate at all. However, women with an IUD generally ovulate. Sometimes a cyst develops in one of the ovaries, which usually disappears on its own.

The vaginal ring (NuvaRing or Annovera) is the pill in a different guise. It's a soft, flexible plastic ring that a woman inserts into her vagina like a tampon. The ring can stay in place for five weeks, during which time it releases a slow, continuous dose of progestin and estrogen. The mucous membrane of the vagina absorbs the hormones, which are then immediately transported to the bloodstream. This means that the total quantity of hormones required to prevent ovulation is less than with an oral contraceptive pill. As with the oral pill, you can decide to skip the seven-day break with NuvaRing by immediately inserting a new ring. In that case, you won't bleed.

The hormones in the ring not only prevent ovulation, but also prevent or deter sperm cells from reaching the uterus. A woman shouldn't be able to feel the ring once it's in place or notice it during sex. However, research shows that 80 percent of men who know in advance that their partner has a vaginal ring say they can feel it. On the other hand, none of the men who weren't told in advance that their partner had a ring could feel it.

Louise (24) *prefers not to use hormonal contraception*

Louise was nineteen when her GP prescribed her the contraceptive pill. She didn't give it too much thought. She had a boyfriend, and the pill was the obvious choice.

A few years later, Louise started doubting her choice. She was feeling depressed and withdrawn more often, and got upset more quickly than before. She'd also gained eighteen pounds. The final straw came when she found herself in tears one lunchtime when she broke the yolk of an egg she'd been frying. What's wrong with me? she wondered. Could it be linked to the pill?

She decided to stop taking the pill and soon noticed a difference. "I felt lighter, although of course I can't say for certain that it was because I'd stopped taking the pill."

For now, Louise prefers not to take the pill. She's pleased her mood has improved and she's lost the weight she gained. She attributes this to having a natural cycle.

But what about contraception? That's a frequent topic of conversation in her friendship group. Louise isn't the only one who struggles with her choice and would rather not take hormones. Perhaps it's a generational thing? Her mother doesn't understand. The pill's great, right? "Perhaps that's how it's seen by the generation for whom the pill meant liberation. It's different for us."

Louise discusses contraception with her boyfriend and he seems understanding. However, she gets the impression that it's mainly her responsibility. She's now considering getting a copper IUD fitted, but she has concerns. She doesn't like the idea of having something inserted in her body and has heard it can cause increased menstrual cramps and heavy periods.

But she'll give the copper IUD a go, and if she doesn't get on with it she might try a hormonal IUD. That's more hormones again, but that IUD works locally and the dose is low. "You have to do something," Louise says. "There's plenty of choice, but the perfect solution doesn't yet exist."

Reliability

Most hormonal methods of contraception are highly reliable. The Pearl Index, developed to express the effectiveness of contraceptives in numbers, gives the contraceptive pill a score of around 0.5, if used correctly. This means that if a hundred women take the pill for a year, 0.5 will have an unwanted pregnancy. Other hormonal contraceptives perform better; the hormonal IUD has a Pearl Index between 0.1 and 0.2, the contraceptive injection scores less than 0.1, and the contraceptive implant has a Pearl Index of zero. Only the NuvaRing scores slightly worse, with a Pearl Index of 1.2.

The differences in reliability can be easily explained. In practice, there's more that can go wrong with the pill or the vaginal ring than with the contraceptive injection, contraceptive implant, or the hormonal IUD. You can forget to take a daily pill or change a NuvaRing. The pill can also be less effective in the event of vomiting or diarrhea. These variations in use bring down the scores of the pill and the vaginal ring. Nevertheless, hormonal methods of contraception easily outperform most non-hormonal methods in terms of reliability. It's therefore unsurprising that many women saw the contraceptive pill as a liberation when it first became available; it was very easy, effective, and gave women control over their fertility. It seemed almost like a magic bullet.

Since then, greater attention has been paid to the downsides of hormonal contraception. Young women in particular are concerned about side effects and look for alternatives that don't contain as many hormones or opt for non-hormonal methods. As women's health writer Kate Muir observes in an article for the *Guardian*, there's "a generation of women, trans and non-binary people binning their pills" after seeing viral TikTok videos declaring birth control "this generation's cigarettes," and after years of lived experience enduring their side effects.

Some women struggle with the idea of taking daily hormones that not only influence their menstrual cycle but perhaps also their mood. They think it's important to have a natural cycle and don't like the thought of ovulation being prevented. Sometimes, hormones that only affect the endometrium and not ovulation—such as the hormonal IUD—can be a good alternative, but some women reject all hormonal methods for various reasons.

Hormone-free contraception

Various alternatives are available for those who would rather not use hormonal contraceptives. Often, these methods have a long history.

One of the best-known contraceptives, the condom, dates back to the sixteenth century. It was initially a type of linen sheath that was pulled over the penis. It was used in brothels and was mainly intended to prevent the spread of sexually transmitted diseases. After rubber was invented, the first condoms came to the market in 1870, while the latex condoms we're familiar with today were invented in the 1920s. In 1984, the female condom—which is inserted into the vagina before intercourse—was invented. With a Pearl Index between 2.6 and 10, it didn't appear very reliable and has never been particularly popular.

In theory, the male condom has a Pearl Index of 2. In reality, though, a lot can go wrong—the condom might break or slip off, for example—which gives it a Pearl Index of 12 in practice. The condom is one of the few contraceptives designed for male genitalia, aside from male sterilization, which is a very definitive method. A contraceptive pill for men never really got off the ground. In a committed relationship where partners trust each other, a "male pill" might be an option, but in most other cases the woman would wonder: How can I be sure that my partner has actually taken his pill?

There are also "natural" methods of birth control, such as the withdrawal method (coitus interruptus), and the rhythm method, a calendar-based method of contraception. In both of these, the man has part of the responsibility. In the withdrawal method, the man completely removes his penis from the vagina before he ejaculates. Couples who use the rhythm method use a calendar to work out when a woman is fertile and do not have penetrative sex on those days. Some couples use a combination of both methods, with withdrawal being used on fertile days.

Otherwise, contraception appears to still mainly be the responsibility of those with a uterus. Hormonal methods of contraception as well as most non-hormonal methods focus on the vagina and the uterus. They need to be administered by women and are generally organized and paid for by women too. That hasn't changed much from the past.

Although people didn't know exactly how fertilization worked for many centuries, experiments to find out began in ancient Greece. Initially, these mainly focused on preventing fertilization by applying liquids or ointments to the vagina prior to intercourse, or thoroughly cleansing the vagina with certain substances after intercourse. The list of substances tried includes resin, honey, various oils, vinegar, brine, and alum. In the fourth century BCE, the Greek philosopher Aristotle advised oiling the vagina with cedar oil or olive oil enriched with lead ointment or incense. Aristotle claimed that the smoother the vagina was on the inside, the lower the chance of conception.

In ancient times, people tried to use magic to prevent pregnancy. Women carried amulets made from a lion's uterus, a cat's liver, a child's tooth, or the insides of a hairy spider. It was also said to help if, after sex, a woman were to stand up immediately, bend over, and quickly try to sneeze.

People also considered the use of clothing and accessories to help prevent pregnancy. In the Middle Ages, for example, the chastity belt came into fashion. It was a dreadful leather and metal device attached to a woman's body by a belt. At the front and back were metal plates that joined together via a hinge in the crotch. There were serrated holes through which feces, urine, and menstrual blood could pass, but which made sexual intercourse impossible. The belt could be locked and a woman's husband or father would keep the key safe. You would expect this phenomenon to have become obsolete a long time ago, but surprisingly enough chastity belts were still used in Hungary as recently as 1933.

Barrier methods

Objects designed to be inserted into the vagina to prevent pregnancy were also developed a very long time ago. They can be regarded as early predecessors to the contraceptive diaphragm we know today. The very first description of something akin to a diaphragm dates back to the fourth century BCE. At the time of the doctor Hippocrates, the advice was to insert a simple wad of wool soaked in honey as far into the vagina as possible to act as a barrier to the sperm. The Italian adventurer Giacomo Casanova (1725–1789) advised his lovers to insert half a carefully dried lemon into the vagina.

The diaphragm or cervical cap we know today might look different, but it's a barrier method just like its early predecessors. The modern diaphragm is a silicone ring that's inserted in such a way that it blocks the cervix so that sperm cells cannot enter the uterus. In order to further reduce the chance of pregnancy, the diaphragm should be used in combination with a spermicide. The diaphragm must be inserted prior to sexual intercourse and remain in place for at least six hours afterwards. After cleaning and drying it, it can be reused. It's important that the diaphragm is the right size, as it has to be able to cover the cervix precisely. A diaphragm usually lasts one or two years and has a Pearl Index between 1 and 3.5.

Another non-hormonal contraceptive is the copper intrauterine device (IUD), which a doctor or nurse inserts into the uterus. The idea of preventing a pregnancy by inserting a device into the uterus was developed by the nineteenth century. The very first IUD was made of ivory or metal. It consisted of a button with a rod that was inserted into the uterus via the vagina. As the object was sometimes expelled during menstruation and often led to infections, it never became popular.

The present-day copper IUD no longer really resembles the early IUDs; it's made from plastic and is the shape of a capital letter T or a horseshoe with copper wire around it. The copper makes the sperm cells inactive and causes a minor inflammatory response in the endometrium, so that fertilization cannot take place. The copper IUD must be inserted by a doctor or nurse and, depending on the quantity of copper used, can remain in place for up to ten years. Once it has been inserted, it can take between seventy-two hours and five days for it to become effective.

A disadvantage of using the copper IUD is that peri- ods can be heavier and more painful. The copper IUD has a thread attached to it that hangs into the top of the vagina via the cervix. This means it can be removed easily, if required.

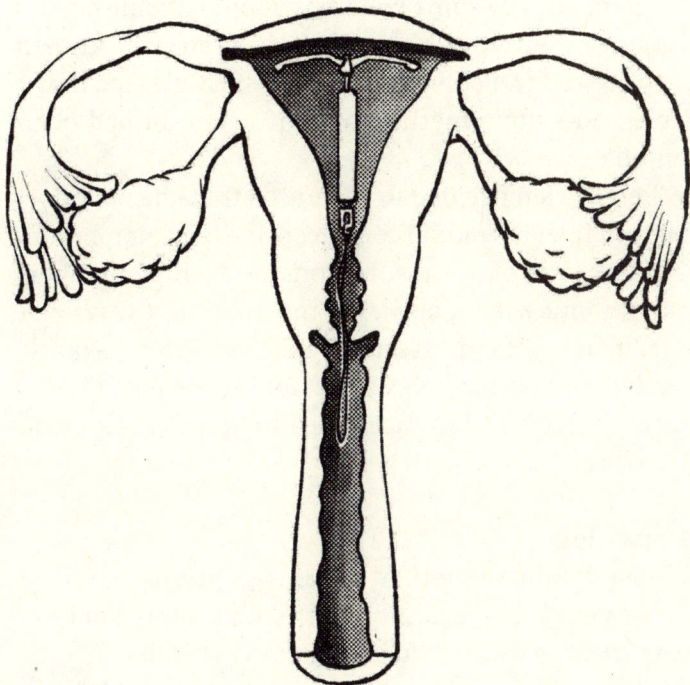

The position of a copper IUD in the uterine cavity.

The copper IUD is one of the most reliable non-hormonal methods of contraception. Depending on the type of copper IUD and the amount of copper used, its Pearl Index is between 0.2 and 2.1.

People have always had concerns about the reliability of contraception. Working on this book has made us think about some of the concerns we had when we were younger. Back then, there was less choice and it wasn't always easy to find the right information; the internet didn't yet exist. There were already some concerns about the side effects of the contraceptive pill, but these weren't as well known as they are today. People were mainly just pleased that a reliable and simple method of contraception had been invented.

When Corien decided in the 1980s that she no longer wanted to take hormonal contraceptives, the diaphragm seemed like one of the best options. She still remembers messing about with spermicide and how she always worried it hadn't worked. At that time, we were constantly reminded that nothing was as reliable as the pill and that every other form of contraceptive brought with it considerable risks.

Sterilization

The most definitive method of contraception is sterilization. It's very reliable, although there's a slim chance of pregnancy afterwards, depending on how proficiently the procedure is carried out. Sterilization can be a good choice if a person doesn't want any (or any further) children and doesn't expect this to change in future. Sterilization can be performed on both men and women.

Male sterilization is easier, less invasive, and cheaper than female sterilization. Male sterilization, or a vasectomy, is carried out by a urologist. During the procedure, the vas deferens tubes that carry sperm from a man's testicles to the penis are severed, usually under local anesthetic. Although this might sound unpleasant, most men can walk out of the hospital half an hour after the procedure. It takes three months for men to become infertile after the procedure. It's still possible to orgasm after sterilization, but the semen a man ejaculates won't contain any sperm cells. For those who regret their decision, it may be possible to reverse the procedure, although this isn't always successful.

The Netherlands has the highest rate of male sterilization in the world, and that rate is on the rise. In Canada and the UK it's a popular choice as well, with around 22 percent of couples choosing vasectomy for birth control. In the US, however, the rate is much lower, about half of what it is in Canada.

If a couple is considering sterilization, for example once the family is complete, the reasoning is often: "The woman has already done a lot with all her periods and pregnancies and she's often taken the pill for a while too. Now it's the man's turn." In countries where contraception is generally considered the woman's responsibility, the percentage of men who have a vasectomy is generally lower. Other factors may contribute too, such as the fact that the procedure may have to be paid out of pocket in countries without government-paid healthcare.

Female sterilization is more complex than male sterilization and can be done in various ways. During the procedure, which is carried out under general anesthetic by laparoscopy near the belly button, the fallopian tubes are blocked or sealed. This is called tubal ligation and is colloquially referred to as "getting your tubes tied." This is done either by putting clips on the fallopian tubes as close to the uterus as possible or by pulling a small section of the fallopian tube through a silicone ring and then clamping it shut.

Another option is to completely remove the two fallopian tubes. The point where the fallopian tube meets the uterus is then cauterized and the whole fallopian tube is separated and removed from the abdomen. The advantage of this procedure is that it may help protect against certain forms of ovarian cancer that seem to originate in the fallopian tube. There are around seven and a half thousand new cases of ovarian cancer each year in the UK, and nineteen thousand in the United States, and it's often discovered late. The disadvantage of removing the fallopian tubes is that the procedure is irreversible.

As long as the fallopian tubes remain, tubal ligation can be reversed if desired. The procedure has a relatively high success rate, but doesn't always work. During the operation, which is performed in an outpatient setting, the part of the fallopian tube that was blocked with the clip or ring is removed and the non-damaged ends are then carefully rejoined.

In 2001, a new sterilization method for women called Essure became available. In this procedure, two flexible metal coils—like the ones found in ballpoint pens—were placed inside the fallopian tubes. This triggered a localized inflammatory reaction, causing the walls of the fallopian tubes with coils inside to stick together and become blocked off. The procedure didn't require anesthesia and could be carried out within half an hour via a hysteroscopy.

The Essure device has since been associated with various side effects and complications, such as abdominal pain and fatigue. As a result, the manufacturer withdrew it from the market in 2017. Many women who are experiencing complications because of Essure want the device removed. This is possible by means of laparoscopy via the abdomen, in which the coils and both fallopian tubes are removed under general anesthetic.

Morning-after pills

Despite the availability of various methods of contraception these days—with or without hormones—there's still a lot that can go wrong. A contraceptive method can fail, be used incorrectly, or be forgotten. Or contraception might not have been used because of a lack of time, will, or care, or in cases of assault. And of course, a couple may have miscalculated, underestimated the risks, or simply buried their heads in the sand about the possible consequences of their actions.

In the past, there wasn't much you could do after unprotected sex other than wait and hope your period would arrive. But now women who fear an unwanted pregnancy can take the morning-after pill. This is explicitly not another method of contraception, but an emergency measure. A morning-after pill differs from an abortion as it prevents a pregnancy as opposed to terminating it.

The morning-after pill has been available since the 1960s. It comes in various forms. A progesterone-based pill works best if taken soon after unprotected sex. This pill prevents or delays ovulation. If ovulation has already taken place, it won't work. If unprotected sex took place more than ten days after the start of the last period, this isn't the best option.

The "normal" contraceptive pill can also be used as a morning-after pill if taken soon enough. Two tablets must be taken twice, twelve hours apart, within forty-eight hours of unprotected sex. In cases where it is too late for either of these options, a medication called ulipristal could be an option. This needs to be taken within five days of unprotected sex. It works by blocking progesterone and making implantation of a potential fertilized egg impossible.

Regardless of the type of morning-after pill used, it is a strong medication that can make a woman feel quite unwell. Women commonly report headaches, nausea, and abdominal pain. In the US, UK, and Canada, the morning-after pill is available over the counter from pharmacies, with the best-known brand in North America being Plan B, and in the UK, Levonelle. The pills are not one hundred percent reliable. It's therefore important to keep an eye on whether the woman's period comes as usual. If not, she may be pregnant and it would be advisable to contact a doctor.

Another option is an IUD. Having an IUD inserted within five days of unprotected sex makes the environment in the uterus unsuitable for implantation of the egg. The advantage is that you can choose to leave the IUD in place as an ongoing method of contraception. A disadvantage is that you need to see a doctor to have the IUD fitted, whereas you can simply pick up the morning-after pill from a pharmacist. There is a very slim chance that the IUD won't work effectively and that a pregnancy occurs.

Gizem (26) *on her abortion*

Gizem had just turned sixteen and was madly in love. One evening, her boyfriend lit candles around the bed and Gizem thought to herself: tonight's the night I'm going to lose my virginity. They were both excited. It was clumsy, painful, and her boyfriend didn't ejaculate. But it was a wonderful night with plenty of kissing and cuddling.

Gizem wasn't initially concerned when her period didn't arrive. It was her boyfriend who bought a pregnancy test. They waited until Gizem's parents weren't at home one evening and did the test together. It turned out she was pregnant. "We cried a lot. It was really intense."

Gizem admits it was naïve of them to have sex without a condom. They hadn't thought about it at all. They also didn't know that pre-ejaculate can contain sperm cells. And they hadn't realized that Gizem was in the middle of her cycle that evening. "I didn't really understand anything about ovulation and periods," she says.

Despite the shock of it all, Gizem immediately knew what she had to do. She was sixteen and wanted to finish school and then go to university. "It felt as if my world was collapsing. This wasn't the plan. I didn't want this."

She didn't tell her family. She wasn't worried that her parents would be angry, but she didn't want to disappoint them. Her family was loving and open, but her Turkish parents didn't talk about sex. Although they knew Gizem had a boyfriend and that they sometimes spent the night together, they had never once mentioned contraception.

She got help and support from her boyfriend's mother and sister. Gizem had the choice between a medical abortion (the abortion pill) and a surgical abortion. She opted for the latter. If she took the pill she would have to wait at home for the bleeding to start, and she didn't want that. Because she was meant to be going on a school trip to Rome—"That was something you spent years looking forward to"—the abortion was postponed for a while. No one in the class knew about it. When they visited a Roman cellar that had mummified babies on display, Gizem broke down. She couldn't stop crying. She finally confided in a friend. "Funnily enough, it was a great trip apart from that," she says. "I had a lot of fun and even managed to forget about everything from time to time."

On the afternoon of the abortion, Gizem and her boyfriend were really tense. The worst part was when Gizem was called through and her boyfriend had to stay behind in the waiting room. "I had to lie down naked on the bed with my legs in the stirrups. It was awful. I felt so embarrassed." She went into a trance-like state after that. The speculum was painful and she remembers feeling a great deal of pressure in her abdomen. Then it was over. "You've had an abortion," she told herself, but it didn't feel real.

That was ten years ago now. Gizem is no longer in a relationship with that boyfriend, but they have a special connection. Gizem sometimes thinks about the child and how old they would be now. She's never regretted her decision. But she does feel a bit guilty that she never told her parents. "It's something I can never change. It's on my track record."

Abortion

If it's too late for emergency contraception, women with an unwanted pregnancy often have no option other than to have an abortion. The Latin term for officially terminating a pregnancy is *abortus provocatus*—*abortus* is the Latin word for "termination" and *provocatus* means "intentional." Amnesty International estimates that globally, one in four pregnancies currently end in an abortion. That's around seventy million abortions a year, or 125,000 a day.

Unwanted pregnancies are as old as time, and the same can be said of attempts to terminate them. Over the centuries, women and those helping them have tried all sorts of ways to end unwanted pregnancies. Their attempts were almost always shrouded in taboo and secrecy. According to the ancient Egyptians, administering hot oils in the vagina would cause an abortion. The Greek doctor Hippocrates, who was strongly opposed to terminating pregnancy, advised a girl who had accidentally gotten pregnant to jump as high as possible seven times so that the semen would come out and fertilization wouldn't be possible.

Over the centuries, in their desperation, women have not only tried strange drinks, oils, ointments, laxatives, and jumping, but have also had others prod around inside them with sharp objects. In the seventeenth century, in secretive back rooms, people used this method to rupture the membrane containing the fetus. A very long nail, knitting needle, crochet needle, or other sharp object was used for the procedure. Sometimes, people attempted to induce an abortion by injecting soapy water or alum into the uterus with a bulb syringe. Many women didn't survive the procedure due to severe bleeding or infections. In those cases, the women who performed the abortions were called "angel-makers."

In countries where abortions are prohibited by law, unsafe abortions are still common today. According to the World Health Organization (WHO), 40 percent of women of a fertile age live in a country where abortion is restricted. Women and girls in these countries who do not have the money or ability to travel to a place where abortion can be performed legally and safely often resort to clandestine methods and the consequences can be fatal. According to Amnesty International, unsafe abortions are the main cause of death in mothers.

The Dutch gynecologist and abortion doctor Rebecca Gomperts, who founded the Women on Waves organization in 1999, is committed to the prevention of unsafe abortions around the world. As a trainee doctor she met women who suffered physically and psychologically due to unwanted pregnancies and lack of access to safe, legal abortion. Gomperts wanted to respond to this need, so decided to offer help from a ship that docks just outside the territorial waters of countries where abortion is prohibited by law. On board, women can have a safe abortion, are given contraceptives, and receive reproductive counselling. Women on Waves has won numerous awards and in 2020 Rebecca Gomperts was listed as one of the hundred most influential women in the world. In 2022, in the wake of *Roe v. Wade* being repealed in the United States, Gomperts began helping women access abortion pills in states where abortion access was banned.

Legal or not

Discussions about abortion can be very heated in many parts of the world and often relate to ethical issues. Should we be allowed to deliberately terminate a pregnancy—the start of a new life? When does an embryo become a person? People have examined these questions throughout history. Thinkers in past centuries believed that an embryo became a person as soon as it started to look like a person and God gave it a spirit. From that moment, the termination of pregnancy was inadmissible as the act would be deemed murder, but the exact timing of that moment left plenty of scope for discussion.

The issues debated in the past aren't too far removed from those today. In 2022, the US Supreme Court struck down *Roe v. Wade*, the landmark decision that had found, in 1973, that abortion access was protected by the Constitution. In response, individual states were able to legislate access to abortion, and laws that had existed prior to 1973 once again came into effect. As of 2024, abortion access was banned or restricted in twenty-one states. For example, in Louisiana, abortion was made illegal in August 2022 (with no exceptions for rape or incest), and abortion providers were criminalized, facing up to fifteen years in prison. In 2024, the governor went even further, passing a law making abortion pills a controlled substance.

According to figures from 2024 from the Center for Reproductive Rights, abortion is currently legal and available on request in seventy-seven countries, which equates to roughly 34 percent of all women of reproductive age. Abortion is completely illegal in twenty-one countries, including the Philippines, Thailand, Dominican Republic, Senegal, and Nicaragua.

Russia was the first country in the world to legalize abortion, in 1920, but Stalin abolished that law in 1936. In the UK, abortion became legal with the Abortion Act of 1967, and in the US, abortion was legal between 1973 and

2022 under *Roe v. Wade*. In Canada, limited abortion was allowed beginning in 1969. In 1988, Canada's laws regarding abortion were struck down altogether, and abortion was treated as a medical procedure governed by provincial medical regulations.

In general, the trend around the world is towards greater access to abortion. For example, in 2021, Mexico ruled that abortion could no longer be considered a crime. And in 2024, France amended its constitution to enshrine abortion access as a right.

Facts about contraception

- The popularity of the pill in the 25–35 age group has dropped significantly in the past years. The IUD and copper IUD have become more popular.

- 18 percent of women of a fertile age globally have been sterilized, compared to 4 percent of men.

- 94 percent of girls and 92 percent of boys use contraception the very first time they have penetrative sex.

- In Japan, the contraceptive pill only came to the market in 1999. It was prohibited before then.

- The very first IUD dates from the nineteenth century.

The path to getting an abortion varies greatly by your country of residence, but the methods remain the same. Prior to the procedure, an ultrasound scan is performed to see if the embryo is visible in the uterus.

There are two methods of abortion. The oldest is called dilation and curettage (D&C). In the past, curettage was performed by inserting a scraper into the uterus via the cervix to "sweep" it from the inside. A teardrop-shaped instrument with blunt upper edges called a curette was used. In a more modern option, vacuum aspiration, a suction tube is used to remove the tissue from inside the uterus. A very thin tube is inserted into the uterus via the cervix, through which the embryo can be removed.

In the early stages of pregnancy (if a woman's period is no more than sixteen days late), an early medical abortion—the term used for abortions induced with medication—is also possible. The so-called "abortion pill" consists of several tablets: a mifepristone tablet is taken first to soften the cervix. After twenty-four to thirty-six hours, four misoprostol tablets are either taken orally or inserted into the vagina (the method varies by country). The tablets cause the uterus to start cramping, which induces a miscarriage. It can take several hours or even several days after inserting misoprostol before the miscarriage starts. In most cases, the miscarriage will be more painful and bleeding will be heavier than a normal period. The blood can also contain clots and tissue. Aside from severe abdominal cramps, side effects also include diarrhea, nausea, dizziness, headaches, chills, and fever.

Since pregnancy tests are now not only more sensitive than they used to be, but also cheap and readily available, women often choose to have an abortion at a very early stage of their pregnancy. In some cases, this might be so early that the embryo is not yet visible on an ultrasound scan. In the US, UK, and Canada, it's possible to get an abortion pill at that time, even without an ultrasound. This is called VEMA, or very early medical abortion.

In the event of a pregnancy longer than thirteen weeks, dilation and evacuation (D&E) is usually performed, as suction curettage is no longer possible. In this procedure, the fetus and placenta are removed using surgical instruments. The procedure is carried out in a clinic, under local or general anesthetic, and takes between ten minutes and half an hour depending on the stage of the pregnancy. Prior to the procedure, tablets are given to soften the cervix.

In the United States, this procedure is politically contentious—right-wing politicians have dubbed it "partial-birth abortion," a term not endorsed by doctors. In sixteen and counting states, the procedure is banned.

Future?

The perfect method of contraception doesn't exist. However advanced it may be, all contraceptives can have side effects or fail. They can be used incorrectly, forgotten, or overlooked. Will that ever change?

It might if the Dutch innovator Peter van de Graaf can find a solution. He's spent the last few years developing a new method of contraception that offers life-long protection against unwanted pregnancy where the contraceptive effect can be stopped as soon as the person wishes. These tiny implants, known as Choice, would be inserted into the fallopian tubes in an outpatient procedure without anesthesia. The implants have a valve that is closed if there is no desire to conceive and open if pregnancy is desired.

This type of on/off switch for fertility, which doesn't require any hormones or active chemical components, might sound too good to be true. Can it actually work? It's not too difficult to block off the fallopian tubes, so doctors expect that part to be successful. After all, that's already what happens with permanent sterilization. The big challenge will be ensuring that the fallopian tubes get through the process unscathed. Will the fallopian tubes still be able to effectively transport sperm cells and the fertilized egg cell once the implants have been inserted? Fallopian tubes are fragile organs and can be easily damaged.

A wide range of technical challenges will also need to be overcome. Work needs to be carried out on a micro level and it will be difficult to choose materials that can be tolerated well and by everyone. However, this is important work. To make progress in the field of contraception, we need to develop new ideas and to turn those ideas into reality.

Isn't it wonderful to think that our daughters or granddaughters may, one day, be able to control their fertility without pills, rings, rods, injections, condoms, ointments, coils, or wire? All they will need is two tiny gates in their fallopian tubes that can open, or close, as desired.

4

The Uterus as a Threat to Health

On fibroids, tumors, and heavy blood loss

"IT'S ALL JUST PART and parcel of it." This phrase comes up time and time again when discussing problems related to the uterus. The fact that uteruses can cramp severely, that they can cause abdominal pain, backache, headaches, and heavy bleeding, and—if you're unlucky—even heavier bleeding, long-term bleeding, or bleeding between periods—it's all part and parcel of it. That all of this can be accompanied by annoyance, sadness, uncertainty, irritation, feelings of shame, and crying fits—well, it's all just part and parcel of it.

That's what mothers tell their daughters, aunts tell their nieces, friends tell each other, and doctors tell their patients. "There's nothing to worry about, it's completely normal, everyone with a uterus goes through it." Then they usually reel off a string of home remedies ranging from hot water bottles, hot baths, herbal teas, and relaxation exercises to a couple of painkillers of your choice off the shelf at your local pharmacy. Plus the often unspoken advice: you'll just have to grin and bear it.

This mentality, which has been shaping attitudes towards so-called "female discomfort" for centuries, is why women often put up with menstrual issues for so long before going to see a doctor. If they go to see a doctor at all, that is. That's a pity, not only because there's often something that can be done about the problems, but also because serious issues often fly under the radar for many years.

Menstrual issues can be severe and although a clear cause can't always be found, sometimes it can. In certain cases and under certain circumstances, the uterus can become a threat to health. Fortunately, that's not the case for everyone. But if it does happen, these are issues that generally won't be resolved with a hot water bottle and painkillers, but which a doctor may be able to treat. In other words, these are issues that need to be taken seriously.

Uterine problems

Spending a morning in Marlies's treatment room gives a good insight into the different ways in which the uterus can present itself as a threat to health. This Tuesday morning we're doing consultations together, both with our white doctor's coats on but each with a different role. Marlies is the doctor carrying out the treatment; Corien is there as a spectator. Every fifteen to thirty minutes, a new patient arrives. Each patient has her own story and her own problems, but the stories all feature the uterus in the main role and often relate to blood loss and pain.

The first patient's problem is benign tumors. She's in her twenties, has dreadlocks, and is wearing a cool jacket. "The fibroids are still there," she sighs. She's referring to the two benign tumors (also known as myomas) in her uterus, which are so large you can feel them through the abdominal wall. One of the myomas is attached to the uterine wall by a stalk and can move around relatively freely. You can sometimes feel it on the left side of the abdomen, then on the right. The patient suffers from severe menstrual cramps and heavy bleeding as a result. Today Marlies performs an ultrasound scan to see exactly how big the two myomas are. The patient will then make an appointment to have them removed in the operating room.

The next patient also suffers from heavy blood loss. She's in her forties and looks nervous. She wraps a thin woolen shawl tightly around her shoulders as she talks. She's here to have a small polyp in her uterus removed. When the endoscope passes through her cervix—that unpleasant moment—it all becomes too much. She explains that she's had a lot of health issues recently. She's usually able to cope with everything, but today she breaks down in tears. She soon regains her composure and undergoes the rest of the procedure with admirable calm.

Then there's a young woman with shoulder-length blonde hair. She's wanted children for a long time but hasn't yet managed to conceive due to adhesions in her uterus. The adhesions—scar tissue that joins internal surfaces in the body—were the result of damage to her endometrium and she recently had them surgically removed. Today Marlies is examining the uterus to see if everything looks good so that she can start IVF treatment. The patient has a long history of gynecological issues. She knows the ropes. Although she's not looking forward to today's hysteroscopy, she's very motivated. This is another step on the long road to fulfilling her one wish: to get pregnant.

The last patient of the morning is suffering from symptoms of perimenopause and is experiencing heavy bleeding. She's wearing a long, baggy blouse and chunky jewelry. She seems a bit flustered when she enters the room. She sits down and immediately starts reeling off a long list of symptoms. As she no longer plans to have children due to her age, endometrial ablation could be a good treatment option. This procedure involves inserting an instrument into the uterus to destroy the uterine lining. The aim is to reduce blood loss during menstruation. Today, Marlies uses an internal ultrasound to examine her uterus and ovaries. The patient will then make an appointment for the procedure to be carried out, which will happen under sedation.

The varied assortment of patients that passes through the doors of the treatment room that Tuesday morning shows just how diverse uterine problems can be and how they affect women of all ages. A uterine polyp, for example, is generally seen in postmenopausal women. A small growth like that in the endometrium doesn't always come with symptoms. Polyps sometimes cause a bloody discharge, bleeding between periods, or postmenopausal bleeding.

Polyps are usually benign, but the incidence of malignant polyps after menopause is around one in twenty. Polyps can be removed by a hysteroscopy. In this procedure, a small hysteroscope is inserted via the cervix to examine the uterine cavity. The polyp can be cut off using special scissors and then removed from the cavity. A pathologist will then examine the polyp to determine whether it is benign or malignant.

Fibroids

The fibroids, or myomas, that affected the first patient in the consulting room are a very common problem. Around thirty percent of women experience them at some point during their fertile years and they are almost always benign, most without symptoms. They are only malignant in one in every ten thousand women.

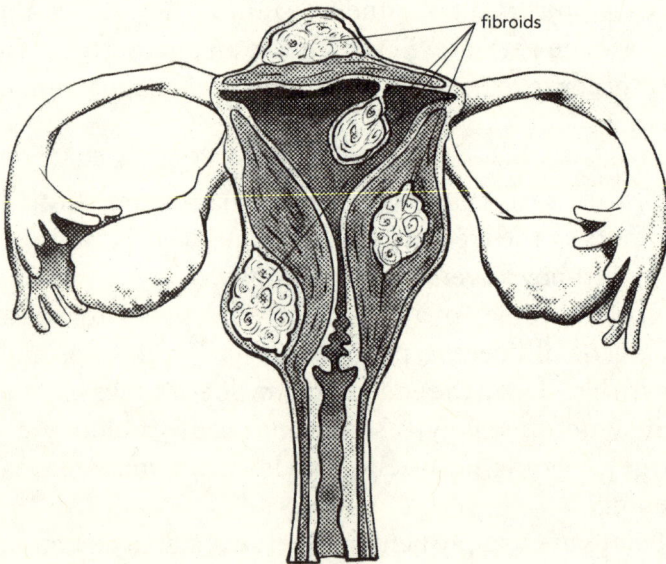

fibroids

Fibroids in various places in the uterus: on the outside, in the uterine muscle, and in the uterine cavity.

Fibroids are growths made up of muscle and fibrous tissue that are formed in the uterine wall. They can develop in women from the first time they get their period, or later in life. Most women with fibroids are between the ages of thirty and fifty. Fibroids usually develop gradually and they may cause pain and blood loss, although they often go unnoticed. After menopause, fibroids shrink and usually no longer cause symptoms. They are most common in Black women.

Not much is known about what causes fibroids, although it's clear that female hormones are involved. We don't know why the muscle cells in the uterine wall sometimes start dividing excessively until a small growth or various growths are formed. The uterine muscle is the only muscle in the body in which these types of tumors can form.

Depending on the size and location of the fibroid, they can cause symptoms of varying severity. Fibroids that develop in the uterine cavity are generally the most problematic and can cause heavy bleeding and menstrual cramps. The uterus tries to get rid of the fibroid by contracting, but it remains firmly attached. The fibroids located in the uterine wall can also cause symptoms, such as excessive blood loss and painful cramps. If a fibroid is on the outside of the uterus, it is less likely to cause bleeding. Fibroids are sometimes attached to a stem on the outside of the uterus and take up lots of space in the abdomen as they grow. This can cause pressure and a heavy feeling in the pelvis.

Fibroids can be seen with the help of a transvaginal ultrasound. If it is suspected that a fibroid is (partially) located in the uterine cavity, a hysteroscopy is often carried out. If the uterus is really large, an MRI scan can give a better picture. Fibroids can be small growths measuring less than half an inch or can be really large, measuring up to five inches in diameter. As the uterus is designed to stretch when a baby is growing inside it, it has no trouble stretching to accommodate fibroids. Occasionally, just like in pregnancy, a uterus can stretch as far as the ribcage as a result of large fibroids in the uterine muscle. This can add a couple of pounds or more of weight.

Irena (47) *on uterine fibroids*

Irena had suffered from heavy periods for years. She was used to them and usually managed fine with plenty of sanitary napkins and tampons. She occasionally experienced leaks at night, despite using thick sanitary napkins, and sometimes had to change her bedding when she woke up. It was annoying, but manageable.

That changed when she was on her way back from a vacation. She'd had abdominal cramps on the plane. Oh no, she'd thought. It can't be my period already, can it? When she landed in Amsterdam, the cramps suddenly became intense, causing her to double over in pain. At the same time, she felt blood between her legs. A big clot came loose, and a huge red stain was immediately visible on her white trousers. How awful! She wanted the ground to swallow her up, but she still had to get through the arrivals hall, and then take the shuttle bus to the long-term car park. It seemed like an endless journey.

"I never, ever want to go through that again," Irena said afterwards, so she decided to make an appointment with her GP. The ultrasound showed a fibroid in her uterus. Irena was referred to the hospital, where the gynecologist confirmed that she had a substantial fibroid. It would be possible to remove the fibroid and Irena would be sedated for the procedure. Irena dreaded it, but it went better than she had expected.

The most important thing was the outcome. Her periods became considerably lighter and the pain reduced. Since the procedure, Irena feels fitter and more energetic. She wants to shout from the rooftops that if women are experiencing serious menstrual issues, they should see their doctor. Something can be done to help.

If fibroids cause symptoms, a gynecologist can treat them. Depending on their location, number, and size, they can be selectively removed. It's possible to treat fibroids without surgical intervention with the help of medication, such as drugs that block the receptor for progesterone. Without progesterone, the fibroid will shrink and cause fewer symptoms. Another medication that is sometimes prescribed causes the pituitary gland to stop producing the hormones LH and FSH. This induces an artificial, temporary menopause, which causes the fibroids to shrink.

The choice of treatment depends on the woman's wishes, the severity of her symptoms, and the size and location of the fibroids. Her age also plays a role and whether she wants to have children. There are various methods of removing fibroids.

Often, a surgical procedure via laparoscopy is possible. In the case of fibroids on the outside of the uterus, the laparoscope is inserted via the abdominal wall. If the fibroids are in the uterine cavity, the procedure can be carried out by means of hysteroscopy. In that case, the hysteroscope is inserted into the uterus via the vagina and the cervix. The fibroid can then be removed with a loop activated by a current, which "peels" the fibroid, or by a morcellator, a surgical instrument that cuts the fibroid into pieces and removes them by suction.

Another treatment option is embolization, which is performed by a radiologist. In this procedure, a thin catheter is inserted through a blood vessel in the groin to block off the blood supply to the fibroid. This causes the fibroid to shrink. Radiologists also sometimes use the magnetic resonance high-intensity focused ultrasound (MR-HIFU) method, which combines an MRI scan with the use of ultrasound equipment. This also causes the fibroid to shrink and reduces symptoms.

The newest method is the Sonata System. This is best suited to fibroids that aren't too large and are located entirely in the uterine wall. In this procedure, the doctor looks into the uterus by using a very thin ultrasound rod inserted via the vagina and the cervix. Thin needles are then inserted into the fibroid. Radiofrequency energy is used to damage the fibroid from the inside, which slowly causes it to shrink and reduces the symptoms.

Another, more radical option—if the fibroids are really large or are causing very severe symptoms—for women who don't want (further) children is removal of the uterus. When the uterus is removed from the body, the fibroids will be removed with it. The procedure, known as a hysterectomy, is usually carried out by laparoscopy via the abdomen. A small uterus can be removed vaginally and a very large uterus can be removed via an incision in the abdomen.

For some women, a hysterectomy is out of the question. They are determined to keep their uterus, even if it's giving them a lot of trouble. To them, the uterus is a vital part of who they are. Other women see it differently and would rather just get rid of it.

Endometriosis

It's estimated that one in ten people who menstruate suffer from endometriosis. That equates to at least 190 million women globally. In people with this condition, tissue similar to the uterine lining (the endometrium) grows outside the uterus, where it settles in and spreads out, causing chronic inflammation and adhesions that result in severe pain.

A characteristic of endometrial tissue is that it's highly regenerative. This makes sense as the endometrium in the uterus needs to keep growing back to ensure a successful pregnancy, making it vital for the survival of the species. In endometriosis, the patches of endometrial tissue outside the uterus behave in the same way, growing under the influence of hormones and bleeding each month. However, unlike blood inside the uterus, this blood has nowhere to go. This causes severe pain.

The abdominal pain caused by endometriosis is often worst during menstruation and sometimes during ovulation. But for some women the pain is almost constant. For them, not a day goes by without abdominal discomfort. Endometriosis can also entail pain during sex, pain when defecating or urinating, heavy bleeding, and fatigue. Every one of these symptoms is shrouded in embarrassment. That's one of the reasons why women often wait so long before going to see a doctor. The impact of endometriosis is huge. The pain can be so extreme that women are left unable or barely able to function. The Flemish film director Ellen Andries, who made the documentary *MyEndo* about life with endometriosis, recalls how she "felt absolutely awful, both physically and mentally, every single day."

In general, endometriosis is benign, but it sometimes behaves very aggressively and can be present in various places. It's often found on or near the uterus: against the back of the uterus, on the cervix, on the fallopian tubes and ovaries, and against the intestines. But it can appear elsewhere too, such as on the peritoneum, bladder, or diaphragm, in the navel, in the scar from a cesarean section, or even in the chest cavity. It can consist of lots of small patches or spots, or it can be more extensive.

It's not easy to detect endometriosis, which is one of the reasons it can take a long time—seven years on average—for it to be diagnosed. The smaller spots are often especially difficult to find. Areas of endometriosis at the base of the cervix can sometimes be seen with the help of a speculum, appearing as blue or brown lumps in the vaginal wall. A transvaginal ultrasound can be used to show endometriosis around the uterus, as well as its presence on the intestines, bladder, and ovaries. An MRI scan can also detect endometriosis.

Areas of endometriosis in the abdomen can be identified by the coagulated blood that has pooled there. They are usually dark in color: brown, black, purplish, or dark blue. In the ovaries, dark-brown fluid-filled cysts known as "chocolate cysts" can form. There is by no means always a

connection between the visible extent of the endometriosis and the severity of symptoms. Sometimes relatively
large areas hardly cause any discomfort, whereas a small
spot right on a nerve can cause very severe pain.

Before effective imaging techniques were available,
it often took much longer to diagnose endometriosis. In
her memoir, *Giving Up the Ghost*, the British author Hilary
Mantel writes that the condition completely ruined her
life as a young woman in the 1970s and 1980s. Suffering
from severe but unexplained pain and bouts of nausea as
a student, she turned to a psychiatrist. "He soon diagnosed
my problem: stress, caused by overambition. This was a
female complaint, one which people believed in, in those
years," Mantel writes. "It was in the nature of educated
young women, it was believed, to be hysterical, neurotic,
difficult and out of control."

Mantel even ended up in a psychiatric institution
before she was diagnosed with an extreme form of endometriosis at the age of twenty-seven. The doctor told her
that the only solution was a major operation. Mantel
agreed—"I'd had more gynecologists than I'd had lovers"—
and her uterus, ovaries, and part of her large intestine were
removed, along with her chances of ever having children.
At the age of twenty-seven, she abruptly went through
menopause.

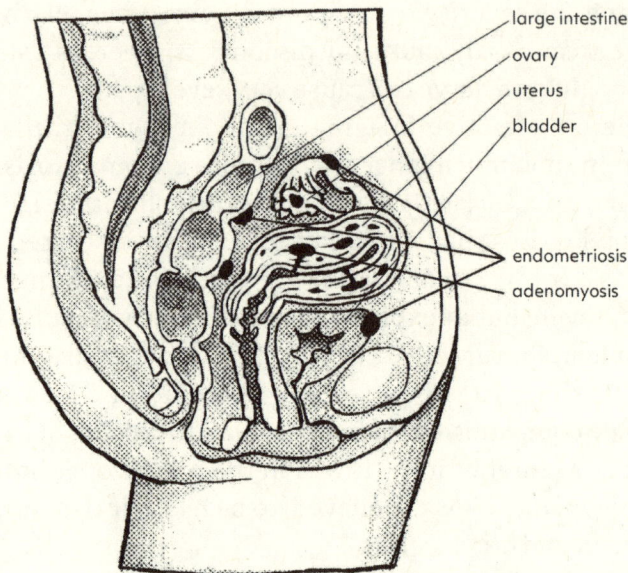

large intestine
ovary
uterus
bladder

endometriosis
adenomyosis

*Cross-section of the abdomen. Areas of endometriosis
can be seen on the uterus, large intestine, and bladder.
Adenomyosis can be seen in the uterine muscle.*

Causes and treatment

There are various theories about the precise causes of
endometriosis. However, one certainty is that it's associated with a hormonal imbalance. Some people believe
that the condition is caused by retrograde menstruation,
when menstrual blood flows back through the fallopian
tubes and ends up in the abdominal cavity. This happens
to everyone who menstruates, and the blood is usually
absorbed without you noticing it. In certain cases, however, it is possible that endometrial cells in the blood seize

the opportunity to establish themselves somewhere in the abdominal cavity, resulting in endometriosis.

Another, more recent, hypothesis points to the stem cells found in the peritoneum. These can sometimes grow or differentiate into endometrial cells, which may lead to endometriosis. Genetic factors are probably also at play. If endometriosis is common in your family, you are more likely to have it yourself.

Endometriosis can't be cured, but treatments can reduce the severity of the symptoms. Some women benefit from alternative therapies, such as homeopathy and acupuncture, although there's no scientific proof of their effectiveness. Researchers in the UK and the US have shown that a special "endometriosis diet" can sometimes help. Based on recommendations to avoid wheat, caffeine, trans fats, red meat, soy, and sugar and to reduce dairy consumption, this diet will not make endometriosis go away, but it could help reduce the associated pain.

Surgical interventions are possible for serious forms of endometriosis, but they're not always successful. Laparoscopic surgery can be performed to remove endometrial tissue. If such tissue is present not only on and around the uterus, but also on the intestine or bladder, a gynecologist will carry out the operation, sometimes together with a general surgeon or urologist. Over time, however, the tissue may return, either in the same place or elsewhere.

Another option is treatment with medication. If drugs are used to suppress menstruation and estrogen levels, the endometrial tissue is unable to continue growing and, in most cases, the symptoms will decrease. There are various ways of doing this: the contraceptive pill (taken without a seven-day break), progesterone tablets, or a hormonal IUD. Medication can also be used to induce an artificial, temporary menopause.

If none of these treatments are effective and the pain is extreme, women may, in exceptional circumstances, turn to the most radical solution: removal of the uterus. That decision can be hard enough for older women who no longer want children. But sometimes the suffering is so great that this radical choice has to be made at a young age.

The American actress and writer Lena Dunham told *Vogue* that she was left with no option but to have a hysterectomy at the age of thirty-one. She had gone through years of extreme pain and tried every treatment in the book, and nine operations hadn't helped. Dunham described how difficult it was to make the decision as it meant giving up her dream of having children. "The pain becomes unbearable. I am delirious with it. [...] With pain like this, I will never be able to be anyone's mother. Even if I could get pregnant, there's nothing I can offer." She explained how her situation became increasingly unsustainable over time. "I check myself into the hospital and announce I am not leaving until they stop this pain or take my uterus. No, really, take her."

The position Dunham found herself in was exceptional. But it's clear that the situation for anyone with endometriosis who wants to have children is complicated. The condition reduces the chance of natural conception, especially if it has damaged the ovaries and fallopian tubes. A woman taking hormones to manage her endometriosis will have to put the treatment on hold when she starts trying to conceive. And when her periods return, her endometriosis and pain usually come back too. For women who wish to have children without going through a lengthy period of severe pain, IVF can be an option.

Sophie (24) *on endometriosis*

Sophie explains how she's had trouble with her periods—or rather, her entire abdomen—for the past six years. Not long after she first started menstruating, her GP prescribed her the pill because her periods were so irregular. At eighteen, Sophie decided to stop taking the pill. She initially felt all right, but looking back now she realizes her periods were already really painful. The bleeding was manageable at first, but the pain in her abdomen was sometimes so severe she couldn't stand up and was forced to crawl around on all fours.

Sophie now knows that her abdominal pain, which could be unbearable, was caused by endometriosis. But a lot happened before she got that diagnosis. To begin with, Sophie's GP referred her to a gastroenterologist due to the pain affecting her entire lower abdomen. He didn't notice anything unusual and suspected she had irritable bowel syndrome, but the pills he prescribed didn't help one bit.

Sophie felt lost; the gastroenterologist couldn't find anything the matter and her GP didn't have a solution either. What was she to do? She felt as if she was always complaining, as if no one understood how awful she was feeling. Sophie could no longer function properly because of the pain. She was having to call in sick more and more often and didn't have much of a social life left.

All Sophie could do was struggle on with her symptoms. The only thing that offered some occasional relief from the severe cramps was a hot water bottle. Sophie has brown patches on her stomach. "Those are burn marks from a hot water bottle that was far too hot," she explains. "Sometimes I was so desperate I tried to mask the pain with more and more heat. The pain was so intense I didn't even realize I was burning myself."

Sophie works in a large clothing store. Her colleague Fatima, who'd joined the same department, noticed how Sophie would sometimes sit in the corner, intensely pale and doubled over in pain. It was a familiar picture. Fatima told Sophie that she had endometriosis and wondered if Sophie might have it too.

Sophie had no idea what endometriosis was, but that night she decided to see if she could find anything out about it online. It wasn't difficult. Astonished, Sophie scrolled through page after page of stories about the condition. She recognized every one of the symptoms. Suddenly she no longer felt quite so alone with her horrendous pain. But she was still reluctant to go back to her GP yet again. Fatima convinced her to make an appointment.

After that, everything happened quickly. Sophie was referred to the gynecologist Fatima had recommended, who confirmed the diagnosis. The ultrasound and scan showed clear areas of endometriosis between her uterus and intestine. The gynecologist suggested fitting an IUD to suppress Sophie's periods. It seemed to help, but the pain didn't disappear entirely.

Sophie explains that endometriosis not only changed her life, but also her personality. She used to be happy, full of energy and up for fun. She liked hanging out with friends and was in a theater group. But the years of pain and lack of answers made her miserable and withdrawn. She gave up acting as she was always having to miss rehearsals. She became depressed. It was awful to be in so much pain without knowing why.

Now that the cause of the pain has finally been found and Sophie has been taken seriously for the first time, she feels relieved. But she's also angry that the condition took so much from her. She's worried about the future because she's read the stories of other people with the same condition, who often need major operations to reduce their pain. For now, she hopes that the IUD will keep doing its job well enough and that she can start enjoying life again. She says, "I hope my story can help other girls and women find the cause of their abdominal pain."

Adenomyosis

The endometrium can also penetrate and proliferate in the muscular wall of the uterus. This results in "corridors" of endometrial tissue in the uterine muscle, which can cause severe cramps, heavy bleeding, and bleeding between periods. This condition is called adenomyosis.

Adenomyosis is difficult to detect. Sometimes it shows up on an ultrasound scan, but an MRI might be needed to diagnose the condition. Like endometriosis, its cause is unknown. Various studies are being carried out to investigate treatments. In some countries, the gynecologist intervenes surgically, cutting away the areas of adenomyosis. This is an operation that can seriously damage the uterine muscle.

The preferred approach is therefore to use medication. Hormonal treatments—the contraceptive pill, progesterone tablets, or a hormonal IUD—are aimed at suppressing menstruation and hopefully reducing symptoms. The idea behind it is that if the production of endometrium in the uterus stops, its growth in the uterine muscle will also be inhibited. Adenomyosis and endometriosis often occur together.

Ovarian cysts

Small or large cysts can form on the ovaries relatively easily. These are follicles that were stimulated but did not ovulate. In many cases, they are harmless fluid-filled sacs that can be left alone if they aren't causing any problems. However, large cysts may cause symptoms such as abdominal pain, a feeling of pressure in the abdomen, and pain during sex. An ultrasound scan is usually able to reveal the contents of the cyst; if the cyst appears black on the abdominal ultrasound, it's likely that it contains fluid.

If the ovarian cyst isn't filled with fluid, it might be a dermoid cyst. These are particularly common in young women. Dermoid cysts can contain sebum, hair, and other types of tissue, such as nails and teeth. Dermoid cysts are present at birth and are the result of abnormal cell growth during fetal development. The cysts grow slowly, but can become relatively large. They are almost always benign. If dermoid cysts have a diameter less than two inches and aren't causing any symptoms, they can be left alone. If they are larger than that, they usually cause symptoms such as abdominal pain, bloating, and menstrual problems. In these cases, the cyst can be removed by laparoscopy; the ovary remains intact.

Larger cysts—fluid-filled cysts as well as dermoid cysts—add weight to the ovary, which can cause the ovary to twist in a process known as ovarian torsion. This can cut off the blood supply to the ovary and cause intense pain, nausea, and vomiting. In this case, immediate surgical intervention is necessary to rotate the ovary back to its correct position with the help of laparoscopy. The cyst is usually removed at a later stage, once circulation has been restored.

In writing this passage, Corien thinks she's starting to understand what happened to her one morning when she was twelve. She can remember it well: the abdominal pain was so intense she almost threw up. Everyone thought it was acute appendicitis and she had an operation that afternoon. The next morning, the surgeon stood by her bed with a mysterious smile on his face. "It wasn't appendicitis," he said. "It was your right ovary. But we removed the appendix anyway, so it won't bother you in future." Corien was twelve and barely knew what an ovary was. She nodded in surprise and didn't ask any questions. She would never have thought that more than forty-five years later, writing a book about the uterus, she would finally see what happened.

Malignancy

Uteruses and ovaries can be affected not only by fundamentally benign conditions such as fibroids, endometriosis, polyps, and cysts, but also by malignant tumors in the case of endometrial (uterine) cancer, cervical cancer, or ovarian cancer.

The endometrium in the uterine cavity can become malignant. In the UK, over nine thousand women are diagnosed with uterine cancer each year, and in the US, around sixty-eight thousand, of which a very small proportion has a hereditary form. In families with a genetic predisposition, uterine cancer usually affects younger women and occurs in combination with bowel cancer. The non-hereditary form of uterine cancer is generally seen in postmenopausal women; the majority of patients are between the ages of fifty-five and eighty.

Uterine cancer develops in the endometrium inside the uterus. The disease usually manifests itself with irregular bleeding, especially in the case of women who have had their last period some time ago. Other signs may include urinary symptoms and blood in the urine, abdominal pain, and severe fatigue. In women with uterine cancer, the endometrium is sometimes as thick as a third of an inch, whereas it would not usually be more than half that thick in postmenopausal women.

Uterine cancer is often discovered at an early stage because it is easy to notice a symptom like blood loss in women who haven't menstruated in some time; in younger women, really abnormal bleeding will also send them to the doctor. This increases the survival rate. If the cancer has not yet spread outside the uterine wall, surgery to remove the uterus and ovaries will suffice. Subsequent treatment will not be necessary and the prognosis is good. If the cancer is already more advanced, additional treatment will be needed post-surgery, such as radiation, chemotherapy, and hormone therapy.

There are various risk factors for uterine cancer. Women who go through menopause after the age of fifty-five are at increased risk. This is because a late menopause means that the woman went through more menstrual cycles without ovulating, so estrogen stimulated the endometrium for longer. This long-term stimulation increases the risk of uterine cancer.

Women who have never been pregnant are also at increased risk because their endometrium never went through the rest periods offered by a pregnancy. Obesity is another risk factor. Fat tissue produces estrogen, which continues to stimulate the endometrium. Finally, long-term use of hormones also increases the risk of uterine cancer. This applies to women who have used hormone supplementation to treat symptoms of menopause and for women who have had hormone therapy after breast cancer.

Cervical cancer is a completely different form of cancer that mustn't be confused with uterine cancer. This cancer doesn't affect the uterine cavity, but just the cervix itself, at the cervical os. Cervical cancer is usually caused by infection with the human papillomavirus, or HPV. Unlike uterine cancer, cervical cancer tends to affect younger women in their fertile years. Most patients are between the ages of thirty-five and fifty. It's a relatively rare disease; around 11,000 people are diagnosed with cervical cancer each year in the United States, 3,000 in the UK, and 1,400 in Canada.

Since screening for cervical cancer was introduced in the Netherlands in 1996, it is much more common for this type of cancer to be detected at an early stage or even in a precancerous form. This is crucial, because cervical cancer is an aggressive form of cancer that rarely has any symptoms at first.

In Canada, the US, and UK, cervical cancer screening usually begins at twenty-one or twenty-five, and is repeated every three to five years with either a Pap test or, increasingly commonly, an HPV test. If precancerous changes to cells are found on the cervical os—the cells are not yet invasive—it can usually be treated relatively easily. A thin layer of the cervical os can be removed to prevent the cancer from developing any further.

Malignant tumors can also develop on the ovaries and in the fallopian tubes. Around one in a hundred women are diagnosed with ovarian cancer. Most patients are over the age of sixty. Ovarian cancer develops deep in the abdomen without the woman realizing. Symptoms usually only occur once the cancer has spread from the ovaries into the abdominal cavity. These symptoms are rather vague and general, ranging from a bloated feeling, pain in the abdomen, lack of appetite, and nausea to a frequent need to urinate. Ovarian cancer can sometimes also cause gastrointestinal symptoms, severe fatigue, and vaginal bleeding.

The exact cause of ovarian cancer is not known, but various factors play a role. For one in seven women, hereditary factors are involved. A genetic predisposition to breast cancer—usually seen in the BRCA1 or BRCA2 gene—is often associated with an increased risk of ovarian cancer. It is believed that the more times a woman has menstruated during her lifetime, the greater her risk of ovarian cancer. That's the case for women who started their menstrual periods when very young and/or had a late menopause. Women who were never pregnant, didn't breastfeed, or only breastfed for a very short time, or women who never took the contraceptive pill are at greater risk too.

If ovarian cancer is discovered early, it's usually by chance—for example, during abdominal surgery performed for another reason. In that case, removing the ovaries and fallopian tubes suffices. If the cancer is only discovered once it has spread, as is often the case, treatment consists of extensive abdominal surgery removing the uterus and ovaries, followed by chemotherapy. Sometimes targeted therapy is possible using medication that targets the cancer cells. Due to its typical late detection, ovarian cancer doesn't have a good prognosis.

Recent research suggests that some sorts of ovarian cancer may start in the fallopian tubes. That's why the fallopian tubes are often also removed during a hysterectomy, leaving just the ovaries.

Facts about disorders of the uterus

- It's estimated that 190 million women worldwide suffer from endometriosis. That's one in ten women of a fertile age.

- It takes an average of seven years to receive an endometriosis diagnosis. Often, this comes after various incorrect diagnoses and treatments.

- 30 percent of women of a fertile age have a myoma or fibroid at any point.

- The uterus is the only muscle in the body in which muscle cells can divide in order to form myomas or fibroids.

- Myomas or fibroids are only malignant in one in ten thousand women.

- Although many millions of women suffer from disorders of the uterus such as endometriosis, adenomyosis, and myomas, limited scientific research into these conditions is carried out. For the most part, their causes remain unknown.

Women's pain

It's clear that the uterus can sometimes be a threat to health. The list of predominantly benign uterine disorders discussed in this chapter is long, and the uterus and ovaries can also be affected by cancer. However, women who—usually after a lengthy period of waiting and hesitating—go to a doctor with problems affecting their uterus don't always feel heard. Their concerns are often classified as psychological or dismissed with the explanation "it's all part and parcel of it."

Problems affecting the uterus, which often involve blood and pain, are still shrouded in shame. People prefer not to talk about them and often fail to take them seriously. Women themselves are often poorly informed and there are still many gaps in the science, too. There is a lack of funding for research into problems affecting the uterus; research funds for gynecology tend to go to fancy fertility treatments or methods of keeping premature babies alive. Of course, these are important fields of research, but the same could be said of serious and common conditions like endometriosis and fibroids, whose cause we don't yet even understand.

The lack of scientific interest in the uterus—unless related to giving birth—appears to be part of a more widespread problem: women's pain isn't always taken seriously. A large-scale study carried out in 2018 analyzed scientific articles from 2001 onwards regarding sex, gender, and pain. It revealed that healthcare professionals were more likely to perceive women in pain as being hysterical, emotional, complaining, or even fabricating their pain. Men with pain were rarely perceived in these terms.

This bias, known as the gender pain gap, has major consequences. American research has shown that female patients are 7 percent less likely than their male counterparts to be taken seriously at the ER. Women are sent away more often than men without having received treatment or medication. The same research showed that it takes women with abdominal pain an average of sixteen minutes longer to receive medicine than men with the same complaint.

The gender pain gap seems to be part of a deep-rooted problem. For BIPOC women, the effect is even more pronounced; it is even less likely for their pain to be taken seriously than for white women. The issue isn't only that doctors and practitioners tend to underestimate the pain experienced by female patients; women do it themselves, too. They learned at an early age to think of their pain—certainly if we're talking about issues relating to the uterus—as something you just have to put up with. Something it's best not to mention.

The results of a study into endometriosis carried out in 2020 found that it takes an average of seven to nine years for women to receive an endometriosis diagnosis. What's more, the diagnosis is missed in 50 percent of cases. All too often, women are told there's nothing wrong with their uterus or that the problem is in their heads, and they are sent home.

Fortunately, this topic is gaining attention. Over the past few years, numerous books, blogs, podcasts, and documentaries about menstrual symptoms and problems relating to the uterus have appeared around the world. Women are starting to find intense pain less acceptable. "Women are made to feel like they are just supposed to 'tough it out' but that is bullshit," the American actress and comedian Amy Schumer posted on Instagram from her hospital bed after surgery for endometriosis that involved removing her uterus. Her post has been viewed 1.6 million times. She calls for women to stop putting up with, hiding, or trivializing their pain. "We have a right to live pain free."

5

The Uterus as a Cocoon

On an extremely expandable
muscle and what grows inside it

PREGNANCY AND CHILDBIRTH ARE the uterus's core business. That's what the organ is designed to do and that's what its processes revolve around. People have known that for a very long time. It was never in any doubt that the organ played a key role as the creator of new life, but how a uterus ended up in a pregnant state was a more complex question. In prehistoric times, people didn't automatically make the connection between intercourse and reproduction. Some ancient people believed that the purpose of sex was to widen the vagina so that the woman was able to give birth when the time came.

The role of sex in reproduction still wasn't clear in Egypt at the beginning of the Common Era. The Egyptians might have suspected that the male seed played a role in fertilization, but they were convinced that conception could happen just as easily via the mouth as via the vagina.

How exactly reproduction worked remained a mystery for many centuries, but at the same time it was completely natural for people to reproduce. If they failed to conceive, they would beg the gods for a pregnancy. Women without children didn't really count, at least according to many stories preserved in the Bible. Fertility and numerous off-spring were paramount.

Early in human history, a woman's entire existence was centered around the cycle of getting pregnant, being pregnant, giving birth, and breastfeeding. As a result, most women rarely menstruated. Few became old enough to go through menopause. We do not know how many children women had on average throughout the ages, but we can assume that large families were the norm. However, we must remember that infant mortality was high and that mothers also had a relatively high chance of death during pregnancy or childbirth.

It's clear that our foremothers' uteruses were almost always occupied with their core business. This has only changed over the past century. The modern uterus dedicates less and less time to its core task. Women in the US, UK, and Canada today give birth to less than two children on average.

This change has been radical. The number of live-born children per woman in Canada, for example, decreased from 6.6 in 1851 to 1.33 in 2022. Life expectancy increased significantly over the same period. These rapid changes have meant that the average length of time during which the uterus is occupied with being pregnant and giving birth has dropped considerably. Nowadays, this takes up no more than 2 percent of the eighty or so years that some-one with a uterus can expect to live.

A great deal has been written about that interesting 2 percent of a woman's life dedicated to pregnancy and birthing. In the context of this book we want to concentrate on the aspects of obstetrics that affect the uterus itself, such as: How does this ingenious organ stretch to five hundred times its original volume? How does it create a cocoon in which a completely new person can grow? And how does it manage to expel that little person from its tight enclosure after nine or so months without too much damage?

Fertilization

Childbirth is the culmination of work started by the uterus around nine months earlier when a sperm cell fertilized an egg cell. Fertilization usually happens in a fallopian tube, along which the egg travels on its way to the uterus. The millions of sperm cells, released into the vagina at a speed of thirty miles per hour when a man ejaculates, travel at a speed of a tenth of an inch per hour as they try to find their way to the uterine cavity and the fallopian tubes. That's a considerable speed for a cell measuring $1/500$ of an inch and is thanks to its long tail, which measures ten times the length of its body. The sperm cells don't move in a very targeted manner, but if they're lucky they pick up chemical signals from the egg. The uterus also helps by making thin, smooth mucus that facilitates the passage of sperm towards the egg.

In humans, the period in which fertilization is possible is rather limited. From the moment the egg is released from the ovary, it only has around twenty-four hours to be fertilized, while sperm cells can survive for a maximum of a week in the fallopian tubes and uterus. Timing is therefore critical and the process is inefficient. Both reproductive partners waste considerable resources. In a process designed to lead to the fusion of one sperm cell with one egg cell, the man supplies as many as 300 million sperm cells and the woman produces hundreds of eggs, of which usually only one matures and is released. Rabbits, for example, go about it in a smarter way; their eggs are only released after sexual intercourse.

The fertilized egg doesn't need much in the first few days after fertilization. As it slowly travels along the fallopian tube towards the uterus, it starts dividing. Upon arrival in the uterine cavity it needs to find a good spot for implantation. In a healthy uterus with a nice, thick endometrium, any spot will do, but the location might not be ideal. If the egg implants itself against the cervix, there is a chance that the placenta ends up in front of the cervix and blocks off the opening. This makes vaginal birth impossible. In the past, this would have usually meant the death of both mother and child, but nowadays this can be seen on the ultrasound and the child can be delivered by cesarean section.

As soon as the fertilized egg has found a spot to implant, a signal is sent to the ovary to start producing the pregnancy hormone human chorionic gonadotropin (hCG). At the same time, the production of progesterone continues, so the endometrium remains intact and menstruation doesn't occur. The absence of menstruation is usually the first sign that a woman is pregnant. The second sign usually comes from a pregnancy test. Accurate, sensitive pregnancy tests are now available from pharmacies or online. All it takes is a few drops of urine on a stick and the result is available within a couple of minutes.

Modern home pregnancy tests have only been around since 1971, but women had ways to test for pregnancy at home as far back as ancient times; these too involved urine. Women were advised to urinate on a bed of wheat, barley, dates, and sand. If the grain sprouted, the woman was pregnant. If the wheat sprouted, she was carrying a boy; if the barley sprouted, she was carrying a girl. It's certainly possible that grain begins to germinate under the influence of pregnancy hormones and that the result of this ancient test was sometimes correct. But this way of determining the baby's sex had no scientific basis.

IVF treatment

With IVF or in vitro fertilization, the egg is fertilized by the sperm cell outside the body. This fertility treatment is suitable in various situations, such as if the fallopian tubes aren't working properly, in the event of unexplained infertility, for people with endometriosis, for couples where both partners have a uterus, or, if egg cells have been frozen, after sterilization that cannot be reversed. IVF can also be performed in the event of abnormal quantity, shape, or motility of the sperm.

Mature egg cells are needed for IVF treatment. First, the woman is given a hormone that stimulates her eggs to mature. The mature eggs are then collected from the ovary using a hollow needle under ultrasound guidance. In the laboratory, the harvested egg cells are put together with the sperm in a Petri dish before being put in an incubator for fertilization.

If fertilization takes place, the embryo can be transferred into the uterus using a thin tube. Then it's a matter of waiting to see if it implants. Any other fertilized eggs can be frozen for future attempts. The chance that an IVF treatment results in pregnancy is around 35 percent.

Transgender men

For a long time, being pregnant and giving birth were believed to be reserved for women, but anyone who has a uterus—including transgender men and nonbinary people—can become pregnant and give birth. The first person known for this was an American trans man named Thomas Beatie, who gave birth in 2008. However, people have questioned whether he was actually the first; other trans men may have given birth before him, but without any publicity.

Some trans men are deterred by pregnancy because of its strong associations with womanhood. In an interview in *Trans* magazine, a man named Jeff says, "After a while I no longer saw pregnancy as something very feminine, but as something very physical. With that thought I revisited my feelings around wanting children. It is a blessing that as a trans man I can become pregnant, whereas cis men can't."

In 2021, Ryan Ramharak, a nonbinary pregnant person, embarked on a legal battle in the Netherlands to have "mother" or "father" replaced with "parent" on a child's birth certificate, a practice that is followed in some other places, such as California, where gender-neutral terms have been available since 2014. Ramharak said, "I don't feel like a woman, but I don't feel like a man either." Ramharak and their male partner decided that they wanted their child to call them "baba" and "papa" respectively. The law has, in fact, been recently amended: as of March 1, 2023, Dutch transgender fathers who give birth to a child can ask the civil servant to register them in the child's birth certificate as the parent from whom the child was born, instead of as the mother. In the United Kingdom, a similar case was recently lost; according to the British court, "mother" is a neutral term.

Shainy (25) *on her ectopic pregnancy*

Shainy has been living with her boyfriend of five years for the past year. They haven't really discussed whether they want children yet. Shainy takes the pill as a contraceptive. She admits she's a bit of a scatterbrain; she often forgets things and is always losing her keys. She sometimes forgets to take the pill too. She usually realizes later the same day and then takes it immediately.

One time, Shainy didn't get a period during the seven-day break and it turned out that she was pregnant. Although the pregnancy wasn't planned, Shainy and her boyfriend were delighted. As they weren't sure how many weeks pregnant she was, an ultrasound was requested. Shainy and her boyfriend were really excited about the appointment. The gynecologist examined her uterus via a vaginal ultrasound, but to their great disappointment there was nothing to see. The gynecologist also checked beside the uterus, in the fallopian tubes, and in the ovaries to see if there might be an ectopic pregnancy, but nothing was found.

Shainy then had blood taken to determine her level of the pregnancy hormone hCG. Because the level was 1480, it was concluded that pregnancy tissue was growing somewhere. As the ultrasound was unable to determine where, the only option was to wait and see. It was a tense time for Shainy and her boyfriend. Shainy was afraid that a pregnancy could be growing in the wrong place.

She was advised to phone the hospital if she had severe abdominal pain or heavy blood loss, even if it was the middle of the night. It was great that she could speak to someone if something happened, but that advice was also concerning.

Two days later she had another blood test and her hCG level had dropped to 1130. Shainy understood that this wasn't a good sign. She had another ultrasound examination, but there was again no indication of a pregnancy. A few days later, just before she was due to return to the hospital, Shainy suddenly got sharp abdominal pains. She had some bleeding too, although not a large amount.

This time, the ultrasound examination revealed some fluid in Shainy's abdominal cavity, which the gynecologist suspected was blood. Her hCG level had since dropped to 680. The gynecologist explained that Shainy had probably had an ectopic pregnancy that had led to a miscarriage. Shainy was told to stay in the hospital for a few hours to see how she was, but the pain subsided on its own. She was allowed to go home. The nurse was really kind and said that the hospital doors were always open if Shainy had any other symptoms. That put her at ease.

After two weeks, the hCG had almost entirely disappeared from Shainy's blood. She didn't have to go to the hospital again. Shainy explains that it took a while for her to come to terms with everything that happened. There was a lot to process: first, she'd unexpectedly gotten pregnant and was happy about it, then she was very anxious and concerned, and then she was relieved that it was over. She now feels sad and is worried about the future.

> She has been advised to phone a hospital promptly in the event of a future pregnancy. They can then do an ultrasound to see whether there's an embryo in the uterine cavity this time.

Ectopic pregnancy

Occasionally, something goes wrong and the fertilized egg doesn't reach the uterus. It can get stuck in the fallopian tube: sometimes in the funnel close to where the egg was released, sometimes further along in the fallopian tube, or in the last part where the fallopian tube meets the uterus. In rare cases, the fertilized egg implants in the ovary. An ectopic pregnancy occurs in 1.5 to 2 percent of pregnancies. The cause isn't always clear. It can be due to ovarian adhesions caused by a previous infection. It's also possible that the embryo itself isn't well developed. However, it can also be simply due to chance.

Since ultrasound equipment became available, ectopic pregnancies have usually been discovered at an early stage. This wasn't always the case. A book detailing the history of medical science in the Netherlands mentions the case of a surgeon named Abraham Cyprianus who was called to the aid of a thirty-two-year-old woman with severe abdominal pain in 1694. The bystanders told him a shocking story: the woman had fallen pregnant twenty months earlier but had never given birth. Her abdomen was hard and swollen, she had a high fever, and she was very weak.

The situation was so acute that Cyprianus decided to perform a cesarean section. Anesthetics hadn't yet been invented; the woman may have been given a few sips of brandy before the incision was made, if she was well enough. The surgeon opened up her abdomen and found a dead embryo in the right fallopian tube that looked almost as if it had been petrified. The patient survived the operation and made a miraculous recovery. Not long afterwards, she became pregnant again and gave birth to healthy twins.

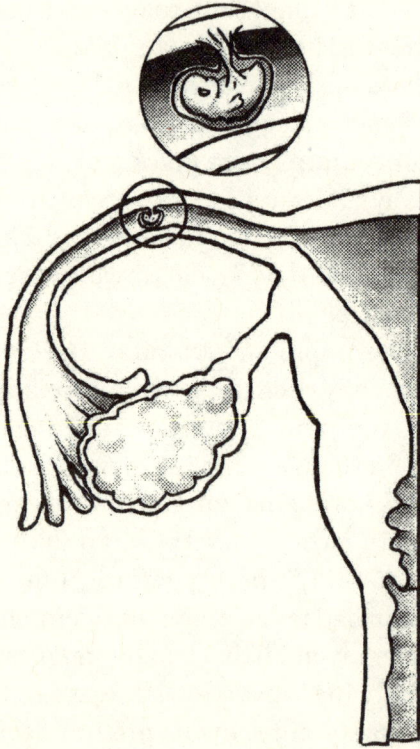

An ectopic pregnancy in the fallopian tube.

These days, surgical intervention for an ectopic pregnancy isn't always necessary. In some cases, the body takes care of it with a miscarriage. In other circumstances, it's possible to treat an ectopic pregnancy with medication. This process requires more time than surgery. If the woman is experiencing acute abdominal pain and if the ultrasound shows a heartbeat somewhere it shouldn't be, there's no time to treat it with medication. In that case, surgery is the only option.

The surgeon usually attempts to save the fallopian tube by making a small incision in it through which to remove the pregnancy. The wall of the fallopian tube repairs itself afterwards. It's sometimes necessary to remove the whole fallopian tube; a person with one fallopian tube should still be able to conceive naturally as long as the fallopian tube is healthy. It's even possible that an egg released on the side of the missing fallopian tube is collected by the funnel-like part of the other fallopian tube.

In some cases, an ectopic pregnancy can leave a woman infertile. This is what happened to the Dutch author Inez van Dullemen during her first pregnancy in the late 1950s. She was on holiday with her husband on a remote Greek island when she became seriously unwell. It turned out to be an ectopic pregnancy. Van Dullemen was operated on urgently and in an improvised fashion. She almost lost her life and was left infertile as a result. She rarely mentioned it for the rest of her life, until she shared this part of her life story in her autobiographical novel *Een schip vol meloenen* (A ship full of melons), which she wrote in her nineties. It was the last novel she wrote.

Ultrasound scans

What happens in the uterus once the fertilized egg implants itself? As soon as the egg has found a good spot, blood vessels are formed in the endometrium; the number of blood vessels continually increases during the first few weeks of pregnancy. They form the start of the placenta. By week twelve, the placenta has become an independent organ. This is when it takes over the production of hormones from the ovaries, which remain mostly inactive throughout the rest of pregnancy. From that point on, the fertilized egg and the placenta develop together.

People have known of the placenta's existence since ancient times. In Latin it was called the *secundina*, or the second thing to be birthed. But the function, nature, and location of the organ remained a mystery until well into the nineteenth century. The lack of understanding was partly because pregnancies generally happened out of sight. A placenta—just like the child—only appeared during childbirth. Until then, the development of the pregnancy could only be followed by feeling and listening. The introduction of the first ultrasound equipment during the 1960s therefore represented a revolutionary change. For the first time, it was possible to see inside the pregnant belly without the need for a knife.

In many countries today, ultrasound scans of the belly are carried out at set times to monitor the pregnancy. For parents-to-be, the first glimpse of their child's beating heart is a special moment. In virtually all modern-day baby books, the pages containing the first baby photos are preceded by a series of black-and-white images that show the development of a speckly white fish into a full-fledged child. Sometimes parents even save ultrasound videos on their phones so they can admire the jerky movements of their unborn baby.

Ultrasound equipment has improved over the years. The images have become sharper and clearer and ultrasound scans are performed on more occasions during a pregnancy. It's customary to have at least two scans— sometimes three or four—during pregnancy in order to assess the baby's development and growth. In addition, parents can go to a private clinic to get ultrasound scans as often as they like during the pregnancy. There's plenty of choice: black-and-white or color; 2D or 3D images. You can even get a video of your unborn child.

Various tests are also available to detect any abnormalities at an early stage. Noninvasive prenatal testing (NIPT) can be done somewhere between week nine and eleven. This uses a pregnant woman's blood to test for certain genetic abnormalities in her unborn child. These are usually chromosomal disorders such as Down syndrome. In addition to NIPT, the nuchal translucency test (an ultrasound to evaluate the neck of the fetus), chorionic villus sampling (a small biopsy of the placenta early in pregnancy), and amniocentesis (sampling some amniotic fluid) are also available to evaluate genetic abnormalities.

Many pregnant women see the twenty-week ultrasound as a milestone. This is the halfway point of the pregnancy. This ultrasound clearly shows any abnormalities, as well as the child's sex. This has given rise to the gender reveal party, a trend that originated in the US and often features pink or blue balloons, streamers, and cakes. Gender reveal parties have come under scrutiny, however, as society's understanding of gender being distinct from sex has grown.

Facts about pregnancy

- Omkari Panwar from India is thought to be the oldest woman to give birth, at the age of seventy. In 2008, she gave birth to twins after undergoing IVF with her seventy-seven-year-old husband.

- Babies start practicing how to cry in the uterus, at around twenty-eight weeks.

- The chance that a mother who has already had twins in one pregnancy gives birth to another set of twins in a following pregnancy is three times as high as for mothers who haven't had twins already.

- Babies can already taste in the uterus. The amniotic fluid they drink from changes flavor depending on what the mother eats.

Multiple pregnancies

Sometimes more than one egg is fertilized at once and this results in a multiple pregnancy. Most of these are twins. The chance of this occurring in the UK is around one in sixty-five, and in the US, three in one hundred. Multiple pregnancies are more common in older mothers. This is because the mother's fertility starts to decline from around the age of thirty-five, which makes it more likely that multiple eggs are released at the same time during ovulation. This is evident when looking at photos of large families from the past. The father and mother can be seen sitting in the middle looking dignified, surrounded by as many as ten children. It's not unusual for the youngest two members of the family to be twins.

A pregnancy with triplets is much rarer than a pregnancy with twins. Triplets occur in around 1 in 1800 pregnancies in the US. Even bigger multiple pregnancies are almost only ever seen after IVF treatment, when multiple fertilized eggs are transferred at the same time. The world record goes to Halima Cissé from Mali, who gave birth to nonuplets via cesarean section in 2021.

In addition to multiple pregnancies caused by two eggs being released at the same time—dizygotic (or fraternal) twins—a fertilized egg may also split to form monozygotic (maternal) twins. As both children come from the same egg and sperm cell, they have almost the same DNA. They will therefore always be the same sex and have a very similar appearance. Monozygotic twins are relatively rare; they only account for 1 in 330 pregnancies.

It's not clear why a fertilized egg sometimes splits. There may be a genetic predisposition to this, as monozygotic twins are prevalent in some families. The earlier in the pregnancy the fertilized egg splits, the greater the chance that the two babies survive. If it splits early, each baby usually has its own amniotic sac, placenta, and blood supply. But other variations are also possible; sometimes the twins share an amniotic sac or placenta and in some cases, they may even share the blood supply. If the egg splits very late, conjoined twins can occur. This is where the two babies' bodies are fused together.

A multiple pregnancy presents a considerable challenge to the uterus. The uterine muscle has to stretch even more than it usually would, which can have consequences for the final stages of the pregnancy. It can be difficult for a muscle that has stretched so much to generate the powerful contractions necessary to push out the baby. A multiple birth is associated with higher risks and premature birth. Whether or not it is possible to give birth to twins vaginally depends on the position of the babies. Many twin pregnancies are born via a cesarean section.

Before ultrasound equipment existed, the birth of multiples often came as a complete surprise. Multiples were therefore considered children for whom God or the gods had a special purpose. There has been much speculation about how many children a woman can give birth to at once. In the early Middle Ages, the Spanish doctor, philosopher, and dentist al-Zahrawi, also known as Albucasis, mentioned a case of "fifteen well-formed children. That's the work of God."

The period of rapid growth

Once the embryo's limbs and organs have been formed in the first half of the pregnancy, the period of rapid growth can commence. The fetus, which doesn't weigh much more than eleven ounces at twenty weeks, will increase its weight fourteenfold in the coming months. The average birth weight of a baby born in the US or UK is about seven pounds. Girls tend to weigh slightly less than boys. In addition to the full-term baby weighing around seven pounds, the uterus also has to carry around the placenta, which weighs between one and two pounds, as well as sixteen to thirty-two ounces of amniotic fluid.

By the end of their pregnancy, women have gained an average of twenty-six pounds, but it's not a problem if they have gained slightly more than that. The increase in weight is not only caused by the growth of the belly, but also because of the breasts getting bigger and fuller as they prepare for breastfeeding. Pregnant women also usually have extra fat tissue and retain more water.

In the meantime, the uterus has stretched as far as possible to accommodate the constantly expanding child and the growing placenta. By the end of the pregnancy, the uterus is around five hundred times bigger than it was at the start. No other muscle in the human body is capable of such a feat. The uterine muscle, which was two inches thick at the beginning of the pregnancy, is not much more than one inch thick by the end. The uterus is now ready for one last, vital effort, when the time is right: it will have to contract to ensure that the fetus can leave the uterus.

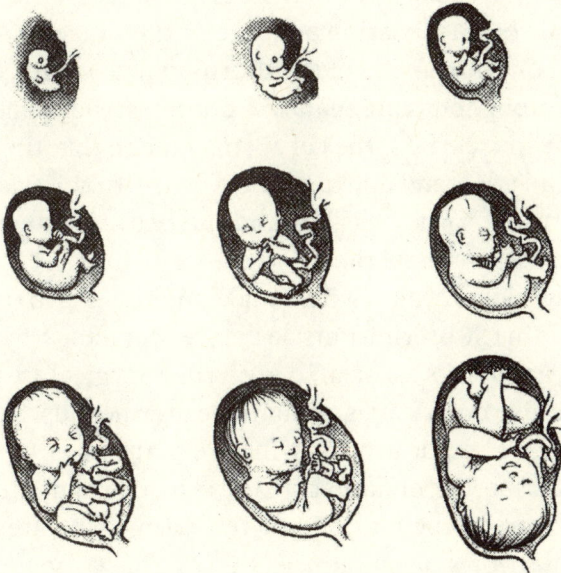

The growth of the baby during pregnancy.

Complications

Pregnancy and childbirth may be the uterus's core business, but successfully completing a pregnancy is a considerable challenge. All sorts of problems can arise along the way.

It's especially difficult for uteruses that weren't functioning too well prior to the pregnancy. Not only do fibroids in the uterus reduce the chance of pregnancy, but they can also cause problems during pregnancy. Fibroids can start growing exceptionally fast during a pregnancy due to hormonal changes, which can lead to pain and a shortage of space in the uterine cavity. However, this isn't common.

Unfortunately, it's difficult to predict whether or not fibroids will cause problems during a pregnancy. A lot depends on the location and size of the fibroids. A small fibroid on the outside of the uterus can go entirely unnoticed throughout a pregnancy. A fibroid on the inside of the uterine muscle near the cervix may mean that the baby's head can't descend due to a lack of space. In this case, vaginal birth won't be possible, and the baby will need to be born by cesarean section.

You might think it would make more sense to treat the fibroids in the uterine muscle before pregnancy, but there are various risks associated with this. Surgical treatment via the abdomen leaves a scar in the uterine muscle, which can cause complications during pregnancy.

Like fibroids, endometriosis can also lead to problems during pregnancy, although the pain associated with endometriosis often reduces during those nine months without any periods. The growing belly can cause adhesions to break off, which can provoke bleeding and pain.

Sometimes that blood needs to be removed via laparoscopy through the abdominal wall.

The microbiome and gynecology

A lot has been written and said about the gut microbiome
in recent years. A healthy microbiome in the digestive system appears to be vital to good health. But what about the
microbiome in the vagina and uterus?

The microbiome is a relatively new term that the molecu-
lar biologist and Nobel Prize winner Joshua Lederberg
coined in 2001. He used the term to refer to the three
pounds of microorganisms—bacteria, viruses, yeasts,
and fungi—that we carry around with us. Most microorganisms can be found in our digestive tract, but the
vagina and uterus also have a microbiome. The microbiome is dynamic and changes according to diet, age, sex,
hormone levels, physical activity, and illness.

Research into microorganisms in the uterus is complex.
It is relatively straightforward to study the vaginal microbiome because the vagina is easy to reach. However, as the
uterus can only be reached via the vagina, bacteria from
the vagina can easily be transferred to the uterus during
the study.

One topic of debate relates to endometriosis. Women
with endometriosis appear to have completely different
bacteria in their abdominal cavity than women without
this condition. This may be an important issue in the
future and could influence the treatment for women with
endometriosis.

It seems likely that the microbiome affects pregnancy
and childbirth. During a vaginal delivery, the baby comes

into contact with the microbiome in the mother's vagina. Researchers think that genetic information from the mother is passed on to the baby in this way. If this is true, it would mean that mothers pass on more genetic material to their child than fathers. But there is not yet conclusive proof of this.

In obstetrics, it's now being debated whether it's necessary for a newborn baby born via cesarean section to be brought into contact with the mother's vaginal discharge. Those in favor think that it can't do any harm and that it might be beneficial. Those against it warn that the vagina could contain dangerous infectious bacteria like Group B strep.

Research into the microbiome is complex and still in its infancy. In future, an individual's microbiome may help us determine targeted treatment for menstrual issues, endometriosis, and much more.

Premature birth

Premature (preterm) birth or threatened preterm labor is a problem that affects 8 to 10 percent of pregnancies in the UK and US, respectively. One of the risk factors is a previous premature birth. Women who have already had a premature birth have a 31 percent chance of another premature birth in a subsequent pregnancy.

In these cases, the uterus starts contracting too early— or far too early—and labor starts even though the baby is not yet ready to be born. A deviation in the shape of the uterus can cause premature labor, but often a clear cause cannot be found. It's not always possible to stop the contractions completely once they have started, but they can often be delayed. Every effort is made for this to happen.

Every additional day that the child spends in the uterus is a bonus.

In the event of threatened premature labor, the mother-to-be is given drugs to prevent contractions as soon as possible in the hope that the contractions will slow down or stop. In the past, these drugs were given for weeks on end, but it has since been established that they only need to be taken for forty-eight hours to be effective. At the same time, babies under thirty-four weeks are given medicine to stimulate the development of their lungs. This improves the baby's chance of survival if labor continues. In a future pregnancy, women are advised to start taking supplemental progesterone at an early stage to prevent the same problem occurring.

Sometimes the cervix also starts opening far too early. This can be caused by previous operations on the uterus or because the cervix has been treated because of an HPV infection. If the cervix starts to open, this increases both the chance of a premature baby and the chance of infection. If this has happened once, it's more likely to happen again. In order to prevent it from happening in a future pregnancy, the most common treatment is a cervical cerclage (or cervical stitch). A cervical cerclage is an operation where a stitch is placed around the cervix to prevent it opening too soon. The procedure is usually carried out vaginally between twelve and twenty-four weeks of pregnancy. A surgeon inserts a speculum into the vagina, holds the cervix, and sews the cervix shut. The sutures are normally taken out at around thirty-six to thirty-seven weeks of pregnancy, unless the woman goes into labor before then.

During a pregnancy, problems relating to the growth and development of the child can arise. These are generally discovered much earlier now than they were in the past, thanks to ultrasound scans. Delayed growth of the baby can have a range of causes. Sometimes it might point to a birth defect or a poorly functioning placenta. It may also be caused by infection.

Preeclampsia can cause delayed growth of the baby. The mother's body reacts to this growth delay by increasing her blood pressure to improve the blood supply to the placenta to aid the baby's growth. A variant of preeclampsia, HELLP (hemolysis, elevated liver enzymes, and low platelets) syndrome is a life-threatening pregnancy complication. Both of these conditions usually occur in the second half of pregnancy.

Preeclampsia is a serious condition for both mother and child. In most cases, attempts are made to delay delivery for as long as possible, but this isn't always an option. If preeclampsia occurs at an early stage of the pregnancy, the longer the baby stays in the uterus the better their chance of survival. If the mother is really unwell, there is often no choice and the baby will have to be delivered by means of an induced labor or a cesarean section.

Kimberley (32) *on the premature birth of her breech baby*

During her pregnancy, Kimberley started feeling her baby more and more in the left side of her belly. She didn't really notice it at first, but it became more obvious to her at around six months. Although she was aware of the baby's position, she didn't pay too much attention to it and the midwife didn't comment on it either.

At week thirty-four, her contractions suddenly started. "Oh no, not yet!" Kimberley began to panic and started hyperventilating, something she'd experienced before. Fortunately, her husband was there and he got her to breathe into a paper bag. It felt strange, but the tingling in her hands disappeared.

The midwife soon arrived. She sent Kimberley straight to the hospital, where the baby's heart was monitored. The baby seemed to be doing fine. In the meantime, Kimberley's contractions had become regular, which meant that she was in labor. An ultrasound was carried out, which revealed that the baby was lying in breech presentation with its head at the upper left side of her belly. The vaginal examination established that she was two centimeters (three-quarters of an inch) dilated.

The gynecologist came to chat with Kimberley and her husband about whether she wanted to give birth vaginally or have a cesarean section. The risks and possibilities of each option were discussed, but Kimberley didn't hear a word. She had already made up her mind; she didn't want to give birth vaginally now. It seemed really scary that her baby's buttocks would come out first. What's more, the contractions were very painful and unpleasant. Kimberley opted for a cesarean section.

From that moment on, everything happened quickly. Kimberley was given an IV drip and told to put on a blue surgical gown. She was taken to the operating room, where she was given an epidural. The pediatrician came and introduced himself, as he would be taking her baby to a special neonatal unit because the baby was six weeks early.

A short while later, a tiny, beautifully pink baby girl was born via cesarean section. Her legs were right up by her chin, but she was doing great. Fortunately, Kimberley was allowed to hold her for a while. The baby girl was placed on her bare chest underneath the surgical drapes. Then she was put into a large transparent incubator and wheeled away to the special unit, while Kimberley stayed behind on the operating table.

The gynecologist peered over the drapes and told Kimberley that she had a partition running through her uterus. That was probably why her daughter had been in breech presentation. There would have been more space for the head at the top of the uterus. What she had felt was exactly how it had been; the baby had been in the left side of the uterus.

Kimberley's daughter was fine. Kimberley went back to the hospital three months later. She was examined by ultrasound and by a laparoscope that was inserted into the uterus via the vagina and the cervix. This confirmed that the uterus was divided into two parts just beyond the cervix, with a thin partition in the middle. In consultation with the gynecologist, the decision was made to leave the partition the way it was. Although it would have been possible to remove it, this wasn't deemed necessary. Kimberley didn't want any more fuss and was pleased that everything had turned out well despite the partition.

Artificial uterus

Premature birth represents a big risk for the baby. With the help of increasingly sophisticated techniques and more advanced incubators, the age at which an infant is considered viable has continually decreased over the past decades and now lies at twenty-four weeks. The incubators that serve as replacement uteruses have become increasingly baby-friendly over time. The babies no longer lie under bright lights, and are put in the fetal position. In the Netherlands, a lot of work is being carried out on the development of something called the Bambi Belt. This is a small belt that monitors various neonatal vital signs so that the baby doesn't need as many injections or bandages. However, starting life in the incubator is still far from ideal. No matter how advanced it may be, the incubator cannot replace the uterus.

Around the world, people are working to develop an artificial uterus in which the baby is surrounded by amniotic fluid and connected to an artificial placenta. One of the main benefits would be that the baby's underdeveloped lungs would not need to function in the artificial uterus. There are various challenges associated with this idea. It's probably technically possible to create an environment that closely resembles the uterus. This is likely to take the form of a balloon-like space filled with amniotic fluid. It also seems possible to develop a heart-lung machine that can function as a placenta.

The biggest challenge will be figuring out how to transfer the infant from the uterus to the artificial uterus without it being exposed to air, which would cause it to start breathing. Some innovators are considering a way of inserting a water-filled chamber into the vagina so that the infant can "slide into" the artificial uterus via the water. Another issue is navigating how to detach the umbilical cord so that it can be quickly attached to the heart-lung machine. Despite these technological question marks, it's expected that the first version of the artificial uterus will be a reality before 2030.

6

The Uterus as an Expelling Force

On the process of giving birth

"THIS IS NOËLLE."

The two of us, both in white coats, are standing at the foot of a bed that's been positioned in the middle of a small hospital room. Grayish light filters into the room through windows with closed net curtains, where it mixes with the bright fluorescent light above the bed.

The sight of the figure on the bed caused a moment of shock. The young woman, pale and motionless, is lying on her back with her arms held stiff beside her body. There's a blood pressure monitor around her right arm. Her white-blonde hair falls around her face in rigid strands. Her colorless lips are half open, as if a scream has just escaped. Both eyes are half closed under a row of unnaturally dense jet-black eyelashes. Just under the two completely symmetrical breasts, there's a gaping hole where the abdomen ought to be.

We're standing beside birthing robot Noëlle in the simulation laboratory at the Máxima Medical Center. She was the first birthing robot in the Netherlands, but has now been replaced by a more modern colleague after years of faithful service. Life-sized birthing simulators like this can be used to train students, to stage emergencies, and to investigate how a team works together.

On a small table beside Noëlle's bed, the plastic baby is ready. Its eyes are tightly shut, and its mouth is fixed in a cry. In the middle of the baby's belly there's a small hole with a thick red border around it. The umbilical cord can be clicked into that hole, which in turn is attached to a fake placenta with bright blue blood vessels against a red background. The umbilical cord is made of transparent plastic so that you can see the intertwined blue and red blood vessels shimmering through it. It looks like the cable from an old-fashioned stereo.

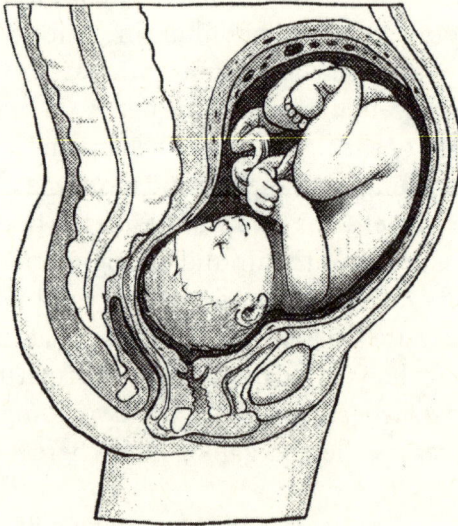

*Full-term pregnancy with the baby
in head-down presentation.*

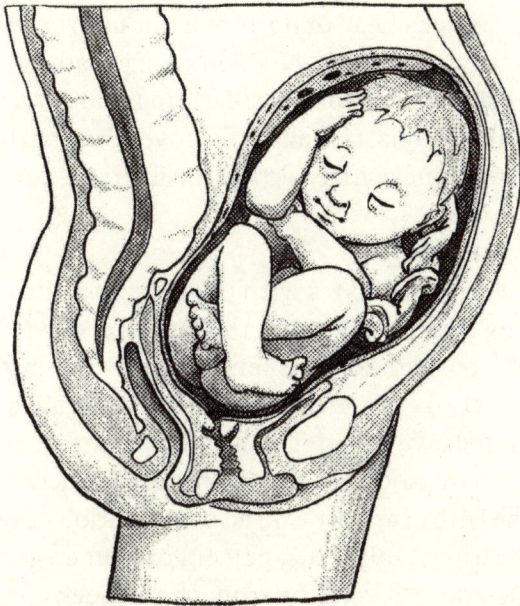

*Full-term pregnancy with
the baby in breech presentation.*

The machinery in Noëlle's abdomen enables the baby to be pushed out in various ways: fast or slow, with difficulty or easily, with or without assistance. If students are practicing a normal birth, the baby is placed into the abdomen with its head lowest down, but a breech birth is also possible. In that case, the bottom is lowest down and that's the part that's born first.

Noëlle's belly is closed with a large skin-colored flap and secured with snap fasteners. Then the power goes on and the enormous belly starts moving to the rhythm of simulated contractions. Students can practice vaginal examinations through the rubbery labia. Can you feel by hand whether a head or bottom is about to appear? As the contractions increase, Noëlle talks, groans, screams, or cries, just like a person would during real labor. The belly surges and the plastic baby doll moves down to the correct position excruciatingly slowly until birth is complete.

The start of labor

For Noëlle, labor starts with the push of a button. But for a heavily pregnant woman, the question of when she will go into labor remains suspenseful until the last moment. From thirty-seven weeks on, the baby is big and strong enough to leave the uterus, so labor can start at any time. The due date calculated in the first months of pregnancy with the help of an ultrasound examination only provides an indication. Only 4 to 5 percent of women give birth on their due date. All births from three weeks before until two weeks after the due date are considered normal.

As pregnancy progresses and the belly becomes large and cumbersome, women start looking forward to the baby's arrival, but sometimes their patience is really put to the test. As a result, there are all sorts of traditional recipes and remedies that are said to induce labor. The ancient Egyptians thought, for example, that peppermint could help. A papyrus text that has been preserved states, "The woman must sit on it with a bare posterior." In ancient Greece, electric eels were placed on the belly of the heavily pregnant woman to trigger contractions. And it was also believed that placing oil-soaked swallow's nests on her loins could help.

There have always been and always will be folk remedies that may or may not be effective. Even today, people have differing views about the effectiveness of a common procedure called a membrane sweep. This is where the healthcare provider puts a gloved finger inside the woman's cervix and separates the amniotic sac from the uterine wall. A membrane sweep can increase the production of hormones called prostaglandins that can encourage labor to start. Many midwives believe that a membrane sweep can be effective, but gynecologists don't always agree, stating that there is not enough scientific evidence to prove that it works.

You can find plenty of other tips and recommendations for inducing labor online. Some people advise foot reflexology and acupuncture, claiming that labor can be induced by stimulating certain zones. Sex is also sometimes advised, as sperm contains prostaglandins that promote labor, and if the woman has an orgasm, the uterus contracts. The oxytocin that's released when the nipples are massaged is said to trigger contractions.

You will find all sorts of nutritional advice too. Some people advise eating pineapple or drinking tonic or bitter lemon, as quinine is said to bring about labor. Spicy food is also said to influence contractions. In her book *Safe Delivery*, midwife Beatrijs Smulders refers to castor oil as "a substance that has been traditionally used by women to induce labor." The idea behind it is that the castor oil causes intestinal cramps, which provoke the uterus to begin cramping too. Finally, exercise like walking or cycling is believed to trigger contractions.

If the contractions still haven't started of their own accord a week after the due date, the risks for the unborn child increase. If the baby's condition remains good, the pregnant person can wait a while longer, but usually they decide to go to hospital for induced labor. Once two weeks have passed, labor induction is almost always strongly advised.

According to a persistent myth, you are more likely to give birth during a full moon. Some people believe that the moon not only influences seas and oceans, but also the human body. The full moon is thought to trigger labor due to its influence on the amniotic fluid. There is no scientific proof of this and if you try to prove the link empirically, you will probably be disappointed.

That's clear from the noticeboard in the obstetrics department at the Máxima Medical Center, which keeps track of all the facility's births. The large felt board has a grid with a box for every day of the year. When new parents leave the hospital, they can put a blue or pink pin in the relevant box. For fun, someone decided to put a gold drawing pin in the board on all the dates when there was a full moon. A glance at the board shows that the boxes with drawing-pin moons don't contain any more blue or pink pins than the "normal" boxes. One of the boxes with a full moon is even completely empty: no children born.

Pregnancy and coronavirus
There has been much debate about COVID-19 and pregnancy. Millions of pregnant women have been vaccinated against coronavirus. The vaccine appears to be safe and effective for pregnant women, for those who have recently given birth, and for women who are breastfeeding. The same applies to parents who are attempting to get pregnant or undergoing fertility treatment.

Pregnant women who haven't been vaccinated and get infected with the COVID-19 virus are at higher risk of getting seriously ill from the infection than non-pregnant women. This applies in particular to women over the age of thirty-five, pregnant women with obesity and/or serious underlying conditions, and pregnant people of non-Western origin. As well, there is evidence that babies whose mothers had COVID while pregnant are at an increased risk of serious respiratory issues after they're born.

At home or in hospital?

In most developed countries, childbirth is medicalized and women generally go to hospital to give birth. In the US, over 98 percent of all births take place in hospital; in the UK, that number is over 97 percent.

Still, the number of home births is increasing. The Mayo Clinic says that some of the reasons people may prefer a home birth are because they would prefer to avoid medical interventions and remain in a comfortable, familiar place. They may also have cultural or religious reasons why they don't want to be in a hospital. It's also possible that they may not be able to afford transportation, or a hospital stay, if they are in a country where this isn't covered.

Some in favor of home birth claim that it's safer giving birth at home than in hospital. Beatrijs Smulders, who is an advocate for home birth in the Netherlands, recently explained this on YouTube. "What I love about home birth is that women are in their own surroundings. They're in charge and they're the ones making all of the decisions," she says. According to Smulders, the birthing woman is most at ease at home, which helps labor progress smoothly. "The safest way of giving birth is where you develop the most powerful contractions."

Corien (59) *on her labors that started off at home and finished in the hospital*

I really wanted to have a home birth. That didn't work out with my first daughter. After having contractions at home for twenty-four hours, there was still minimal progress. That's why the midwife suggested going to hospital to "finish off" my labor there. At that point, I was okay with it. I was exhausted from all those hours of labor and from the repeated disappointing news that I wasn't getting anywhere with dilation.

When I was pregnant the second time, I really hoped I'd be able to give birth in my own surroundings. My mother had often talked about her wonderful home birth, which had gone so much more smoothly than her previous births in the hospital. That's what I wanted too.

In the meantime, I was already a week overdue. According to the midwife, the baby was doing great and there was no reason to induce labor. She had done a membrane sweep in the hope it would get things going. I'd cycled home afterwards, full of anticipation. Now it was sure to start! But two days later nothing had happened, apart from a few Braxton-Hicks contractions.

I went to bed early that night. With my huge belly, I was just trying to find a comfortable position to lie in when it suddenly felt as if something burst in my lower abdomen. I'd never felt anything like it before, but immediately knew what it was: my water had broken. As I made my way to the toilet, I felt the amniotic fluid trickle out of me. Before I'd sat down, I felt an enormous contraction.

The midwife arrived quickly. She saw that the contractions were intense and studied the amniotic fluid in a glass. "Look, it's a bit cloudy," she said. "We'd better go to hospital to get it checked out." The contractions had become so painful I didn't really take in what she was saying. I had to focus all my energy on riding out the contractions and it didn't make any difference if I was doing that at home or in the hospital. The only thing that mattered was that I would give birth to a healthy child.

At the hospital I learned that the amniotic fluid wasn't too concerning and that the baby was doing well. However, the contractions had become even more intense and closer together. It was as if I was in a different world. The surroundings didn't matter, I lost track of time, I forgot about the candles and music I'd envisioned for my home birth. Nothing external seemed important anymore.

Our second daughter was born in the middle of the night. A very healthy girl weighing just over eight pounds. She immediately started breastfeeding, while I enjoyed eating toast with sprinkles in my hospital bed, a Dutch tradition when a baby is born. We were able to leave the hospital with our newborn baby at half past seven the following morning. It was Sunday and the road was just as dark and quiet as when we'd left a few hours earlier. Everyone was still asleep. We'd only been away for one night, but we came home with a daughter!

Hospital births are a relatively new phenomenon. Until the industrial revolution, all women gave birth at home. Childbirth was a women's matter that didn't involve men. For many centuries, obstetrics was a purely empirical subject. Mothers, aunts, older sisters, or female neighbors with experience would offer help during childbirth. Professional midwives who learned through practice have existed since ancient times. These were generally older women who had already given birth themselves and acted based on experience. Midwives were regarded highly, but hadn't usually been formally trained.

From the sixteenth century onwards, booklets about obstetrics started to become available in Europe in various languages. In the following centuries, the first official midwife training courses were established. Childbirth still took place at home and was usually overseen by a midwife. A (male) surgeon was only asked to assist in complex cases. The diaries that the famous Dutch midwife Catharina Schrader kept between 1693 and 1745 reveal that she carried out her profession until the age of ninety. Her notes fill more than five hundred folio sheets in which she describes every birth in telegram style. During her long working life, Schrader supervised more than four thousand births; the baby survived in all but ninety cases.

Effacement

The very start of labor is shrouded in mystery. In 4 BCE, Hippocrates claimed that the child itself initiates birth, and he likened delivery to the process by which a chick hatches out of an egg. For many centuries, this was the generally accepted view.

It has since been ascertained that the initiation of labor is the result of a complex interplay between mother and child. The child gives the first signal that the time has come for delivery around five weeks before birth. This causes the uterus to start preparing itself—for example, through practice contractions known as Braxton-Hicks. Many women experience these as a tightening in the abdomen.

The powerful contraction of the uterus causes the abdomen to turn into a hard, tight ball.

Despite the vast amount of research carried out in this field, it is still unclear what exactly sets the irreversible process in motion that leads to delivery. We do know that the uterus plays a main role. The child doesn't force its way out like Hippocrates believed. Instead, the uterus pushes the child out using vigorous contractions.

The very first contractions, initiated by the hormone oxytocin, usually announce themselves tentatively. They aren't normally very powerful or regular at first. They cause the cervix to soften and thin and the "spout" of the cervix (the cervical os) to shorten and soften until it disappears completely. This process is called effacement. Sometimes the uterus has already carried out preliminary work in the weeks leading up to delivery with the help of Braxton-Hicks contractions. A midwife or gynecologist can feel that; the cervix feels a bit softer and riper. But that's by no means always the case. Especially with a first birth, it can take some time for effacement to be complete and for everything to be ready for the dilation contractions, the contractions that cause the cervix to open.

If this takes a really long time, help is sometimes needed. There are tablets that can be inserted vaginally to help the cervix ripen. A small balloon can also be used, which is inserted into the cervix with a thin catheter via the vagina and then carefully inflated. Sometimes it produces results quickly, but it may still take one or two days before the cervix is completely soft and ready to open.

Once the cervix is ripe and soft and has already started to open, whoever is supervising the birth can break the water in an attempt to stimulate contractions. Delivery can also start with the water breaking spontaneously. Delivery has to happen relatively soon after the water has broken. If there hasn't been much progress after twenty-four hours, the pregnant woman will have to go to hospital to be induced. The risk of infection increases once the water has broken, so it isn't a good idea to wait any longer.

Dilation

After effacement, dilation contractions begin, which are intended to open the cervix. In order to give birth to the child, the cervix has to be fully dilated to ten centimeters (four inches). This can take some time. On average, the cervix dilates by one centimeter every hour, but the process can sometimes take twelve or more hours, especially with a first birth. However, dilation may also take place quickly and intensely. That's seen most often in women who have previously given birth.

For the first couple of centimeters (or one inch) of dilation, the contractions are usually irregular and tolerable. They become more regular, powerful, and painful after that. They come every few minutes and can last between sixty and ninety seconds. A good dilation contraction comes slowly, grows, becomes more and more powerful, and then slowly subsides. Sometimes the process takes place so quickly that you're unable to distinguish between the contractions. Full dilation is achieved when the edge of the cervix can no longer be felt. The expulsion stage can now commence.

Contractions are intense not only for the mother; they also impact the child in its tight enclosure. That's why the baby's heart rate is constantly monitored during labor. The midwife checks the heart rate at regular intervals with a small hand-held ultrasound device called a Doppler fetal monitor. In the past a type of horn was used for this, which was placed on the belly to listen to the heart rate. In hospital, the baby's heart rate is usually continually monitored via internal cardiotocography (CTG). This involves a small wire being placed on the baby's head to monitor the fetal heart rate, which should be approximately 120 beats per minute during labor. The heart rate regularly jumps around a bit, which creates a varied picture.

If a woman gives birth in hospital, her contractions are often monitored by sensors held in place with elastic bands around her abdomen. A special bandage is currently being developed to monitor contractions more easily. This will give the mother greater freedom of movement as she will not have to stay in the hospital bed. Furthermore, the elastic bands, which sometimes pinch the skin, will no longer be necessary to keep the sensors in place.

When dilation takes a long time, it's exhausting for the mother. Stress and anxiety negatively impact dilation. That's why it's important that the birthing parent feels safe and supported. Encouraging and reassuring words from people can play a big role. Warmth, both physical and emotional, is important. A hot water bottle can sometimes help with the pain, and women are often advised to take a warm shower or bath. In addition, breathing techniques learned in a pregnancy course can come in handy when it comes to handling the contractions.

Sometimes, nature can use a helping hand during the dilation phase. In this case, the mother will have to go to hospital, where she will receive an IV drip containing medication to stimulate the contractions. This is induced labor. Sometimes this is necessary due to the condition of the baby, the condition of the mother, or because the woman is very overdue. But induction may be necessary if the contractions aren't powerful enough, the process doesn't progress, or the mother gets too exhausted.

At first, a low dosage of oxytocin is usually administered via an IV drip. This stimulates the uterus to contract. In some cases the uterus then immediately starts contracting itself, meaning that the oxytocin is barely needed, or no longer needed. If that isn't the case, the oxytocin is gradually increased until there are good contractions and dilation starts to progress.

Contractions can be really unpleasant. They stretch the cervix in an intense way and cause severe abdominal pain. Like menstrual cramps, contractions can sometimes be felt not only in the abdomen but also in the lower back or thighs. That's because the uterus is attached to the pelvis with ligaments. When the uterus powerfully contracts, these ligaments are pulled, which can cause severe pain in the groin, legs, or back. The severity of pain during childbirth is described in the first book of the Bible, Genesis, where God says to the woman: "I will greatly multiply thy sorrow and thy conception. In sorrow thou shalt bring forth children."

In the past, the only option women had to help subdue pain during childbirth was often a few sips of brandy (though opium has been used to relieve pain during childbirth in some places for thousands of years). There are better options now.

Various forms of pain relief are available in hospital, such as an IV drip containing remifentanil. This drug has a morphine-like effect and works within a couple of minutes. The woman giving birth determines the dosage and can give herself more pain relief at the push of a button. A midwife or doctor should stay with the woman to make sure that she responds well to the medication.

A more common option is an epidural. This provides localized, long-lasting anesthesia. The anesthetist injects it just outside the sac of fluid around the spinal cord to ensure that the muscle fibers and nerves that transmit pain are anesthetized, but not the sensory nerves. The woman no longer feels any pain in her lower body. She can also feel a sense of pressure in her lower body. An epidural takes fifteen minutes to take effect and can be topped up if pain increases.

A spinal block, given in the spinal cord, is the most commonly used method of anesthesia for a planned cesarean section. It works much more quickly than an epidural, but it can't be topped up. The advantage of a spinal block for a cesarean section as opposed to general anesthetic is that the spinal block works locally so that none of the pain medication is passed on to the baby. And in contrast to a general anesthetic where a woman is unconscious, with a spinal block she can experience the birth with her partner beside her.

Chiara (34) *on her vacuum delivery*

Ten years ago, Chiara moved to the Netherlands from Italy after studying applied physics in Milan. She'd landed a job at a technical university and found love in the Netherlands. Chiara soon became pregnant with her first child. Everything was going well, and as is customary in the Netherlands, Chiara was being monitored by a midwife. At thirty-eight weeks and two days, Chiara's water broke, but unfortunately her contractions didn't start of their own accord. The midwife therefore sent Chiara to hospital so that her labor could be supervised there.

However, on the way to hospital, Chiara's uterus suddenly started contracting. Chiara felt the pain mainly in her back. In the obstetrics department a recording of the baby's heartbeat was made, which indicated that the baby was doing fine. However, the contractions were very irregular. Chiara was carefully examined to determine how dilated she was. She appeared to be only one centimeter (less than half an inch) dilated even though her cervix had already become very short.

They waited a while to see whether the contractions would get stronger and more regular on their own. Chiara's contractions were intense and the pain in her back was sometimes unbearable. After two long hours of contractions, she was disappointed to discover that she was still only one centimeter dilated. Help, Chiara thought. How long is this going to take? I'll never keep it up.

She was advised to take a bath in the hope that the warm water would help alleviate the pain. It felt good, but after twenty minutes the pain in her back was so intense that she didn't know what to do. How was she meant to get out of the bath? Several people had to help lift her out. She was examined again to see how far dilated she was. She was still only at one centimeter, even though her cervix was completely effaced.

What now? Chiara was exhausted and disheartened. The decision was made to help the uterus along a bit by giving Chiara an IV drip containing drugs to stimulate contractions and to administer an epidural for the pain. What a relief! The severe pain in her back finally disappeared. Now she could relax a bit.

Dilation continued to progress slowly, but she finally reached the required ten centimeters (four inches). The baby was lying with its nose pointing up, in the "sunny side up" position. Chiara was asked to lie on her left side in the hope that the baby might turn around so its nose would point down, but it didn't. In the meantime, the uterus was still being stimulated thanks to an IV drip.

Then it was finally time to start pushing. Everyone in the room encouraged her. Chiara found it difficult, as she was too numb to push. After pushing for an hour, the baby still hadn't been born. Then the doctor said they would give her a helping hand—with a suction cup. Various people entered the room, all women, one of whom introduced herself as the gynecologist. Fortunately, the doctor who had been there all along stayed to provide explanations and instructions. At least that was a familiar face.

A small suction cup was placed on the child's head. Chiara had to keep pushing hard during the contractions, and the doctor helped by pulling. Suddenly she felt a lot of pressure and pain. It was completely different to what she had felt up to that point. And then her baby boy, who had turned around at the very last minute, was born face down after all.

Chiara was delighted, but suddenly felt very unwell and incredibly dizzy. She had lost a lot of blood very quickly. Her placenta was still inside and her uterus still wasn't contracting well on its own. Various people streamed back into the room. Chiara was given another IV drip.

Chiara struggles to remember what happened next. All she knows is that she was rushed into the operating room. When she woke up, she didn't know where she was until she saw her husband with the baby in his arms. Chiara had lost almost two quarts of blood. After her placenta was removed in the operating room, they had given her more medication to encourage the uterus to contract and shrink. That helped.

Chiara has been advised to give birth in hospital in the event of a future pregnancy. She's happy with that. She hadn't planned to give birth at home anyway. Home births are uncommon in Italy.

Giving birth

Once the cervix is dilated, delivery can begin. Together, the uterus and vagina form a birth canal that the baby must travel along on its way to the outside world. The uterus takes care of contractions, causing the baby to be slowly pushed out with considerable effort on the part of the mother. This works best if the baby is in the right position, with its head down. The head is the biggest part of the body and therefore the most difficult to birth. As the fontanelles—soft spots in the skull—are not yet closed, the skull can be deformed a bit as it travels along the birth canal. If the baby's nose is pointing down, delivery goes most smoothly as the smallest part of the head exits first. A baby with its nose pointing up means a wider part of the head exits first, which can make delivery a bit more difficult.

When a baby is delivered with its lower body going first, it is called a breech birth. During pregnancy, all babies are oriented with their bottom down every now and then. In the months when there is still sufficient space in the uterus, the baby regularly turns around. At twenty-five weeks, 30 to 40 percent of babies are in breech position. By thirty-two weeks, that figure has dropped to 10 percent, and by the time they are born, only 3 percent of babies are in breech position. That's because the head is the heaviest part, so most babies automatically drop down into the pelvis with their head down towards the end of the pregnancy. If that doesn't happen, it's possible to try and turn the baby in the weeks prior to birth. It's not a pleasant sensation for the mother, but it often works.

As there are more risks associated with breech birth, women in this situation often decide to have a cesarean section, but it's sometimes possible to have a vaginal birth anyway. If the baby is in complete breech position, the legs are born first. If the bottom comes out first, it's called an incomplete breech. These days, breech presentation is a reason to give birth in hospital under the supervision of a gynecologist.

Transverse presentation, where the baby lies sideways across the uterus, occurs less often than breech presentation. If the baby is in transverse position, attempts are also made to turn the baby before birth. If that isn't successful, a cesarean section is the only option, as it is not possible to give birth vaginally to a baby in the transverse position.

It's possible for a birthing parent to give birth in various positions. The lying or half-lying position with raised legs that is customary in many hospitals only came into fashion when instruments were developed in the seventeenth century to help labor that wasn't advancing. If a doctor is using tools to help deliver a baby, the mother-to-be needs to be lying down. Before these instruments were developed, women not only gave birth lying down, but also crouching, kneeling, or standing. The advantage of these positions is that gravity offers a helping hand, although that's also a risk: in a vertical position, there's a greater chance of tearing.

The earliest reports of a birthing chair or birthing stool date from the second century CE at the time of the Greek doctor Soranus of Ephesus. If such a chair wasn't available, the woman giving birth would sit on another woman's lap. This woman, who would sit with her legs wide apart, would have to be strong enough to bear the mother's weight and keep her still. She would often be assisted by attendants on each side who would support the woman under her arms. Soranus of Ephesus was the first doctor to advise supporting the birthing woman's perineum with a linen pad to prevent the skin between the vagina and anus from tearing. The midwife would sit on the ground in front of the birthing woman and bring the child into the world by feeling under the wide skirts of the mother-to-be.

Marlies (64) *on being pregnant while training to be a gynecologist*

In my third year of training to be a gynecologist, I suddenly found out I was pregnant. I realized I hadn't been careful with contraception over the past few months. The pregnancy came as a big surprise, and it was unusual for a trainee to be pregnant. In those days—the late 1980s—female gynecologists were still in the minority, never mind pregnant gynecologists. But I was delighted.

Although it was a low-risk, healthy pregnancy, I asked one of the gynecologists from the group I was training with to oversee my pregnancy. I really wanted to be under the supervision of someone I fully trusted and who would tell me if I needed to slow down. Fortunately, the pregnancy was uncomplicated, although I had an enormous bump. I gained a total of fifty-five pounds. Nevertheless, I carried on working until I went on maternity leave at thirty-four weeks.

My contractions started at bang on forty weeks and we left for the hospital in the afternoon. The contractions gradually got stronger, but dilation progressed slowly. In those days, an epidural or any other form of pain relief wasn't offered, so it was a matter of gritting my teeth. I stood up for a bit, then had a lie-down, then rocked back and forth. I did everything to ride out the contractions. They seemed to go on forever.

At ten that evening my husband was sent away to the waiting room. I think there was a football match on the television. He was able to relax and recharge a bit in preparation for what was to come. It was a really special evening and night. I was supported by my husband, the nurses, but above all by my gynecologist. With lots of love, patience, advice, hot water bottles, and words of encouragement I finally became fully dilated.

After a very strong push my son was born, weighing almost ten pounds. My colleagues shrieked: it could have been two babies! We felt incredibly happy.

Assisted birth

In the past, labor that failed to progress—where the woman experienced endless, painful contractions without the baby being born—was a worst-case scenario. If the situation seemed hopeless after some time, and the mother was weak and exhausted and lost a lot of blood, the advice until the sixteenth century was to let nature take its course. Or to leave it in God's hands, which amounted to the same thing. In those cases, neither mother nor child would survive.

These days, we have tools that can help bring a labor that's failing to progress to a positive end. Before the seventeenth century, instruments had been used to remove a dead baby from the uterus—with varying levels of success—but the idea that you could use an instrument to help bring a living child into the world only arose when obstetrics began to develop into its own branch of science.

Early attempts to develop these instruments weren't all equally successful. The "iron screw" designed to stretch the cervix and pull the pelvic bones apart in the event of incomplete dilation is a notorious example. It was dubbed the "mother murderer."

The idea of using forceps to guide the baby's head out if labor is failing to progress originates from the seventeenth century. The first forceps were known as the "secret instrument" for a long time. The developers didn't want to make their model public as they hoped to profit from it. It was only one and a half centuries later that the design was released. Hinged forceps were later developed, which set an example for the forceps we know today.

In Canada and the UK, up to 15 percent of all births are assisted births; in the US, the number is significantly lower at 3 percent. A gynecologist performs an assisted birth in the hospital with the help of either forceps or a vacuum device. Assisted births are most common in first pregnancies.

Traditional forceps are still available in every hospital, but they are rarely used these days. They consist of a handle with two oblong scoops attached. The scoops close around the baby's head. By pulling the handle of the forceps, the gynecologist can help guide the baby's head out during a pushing contraction. As use of forceps often results in damage to the mother's pelvic muscles, the vacuum pump is generally the preferred choice.

Like forceps, the vacuum pump also has a long history. The first attempt to place a glass cup on the baby's head for vacuum delivery dates from 1705. It wasn't a success. The first real vacuum pump from 1849 wasn't well received either. The equipment had a rubber suction cup that could be used to create a vacuum by hand. A medical historian from the nineteenth century summarized the invention as "one of the many errors in which the history of the forceps is so abundant."

There are two versions of the modern vacuum pump. The traditional pump has a metal or plastic cup with a diameter of two inches that is placed on the child's head. The cup is attached to an electrically driven pump via a hose. A smaller vacuum extractor called the Kiwi is also available; in this version, the cup is attached to a hand-held pump.

Both vacuum pumps are based on the same principle. As soon as the cup is secured on the baby's head, the obstetrician can pull every time the mother has a pushing contraction. In many cases, an assisted birth is an intense, painful experience for the mother. It's not easy for the baby either. Babies born with the help of a vacuum pump often have a bruise on their head where the cup was attached. They are given acetaminophen for the first few days after birth to reduce the pain. The swelling usually goes down after a couple of days.

Habibe (33) *on her cesarean section*

Habibe and her husband wanted a third child. They already had a six-year-old son and a three-year-old daughter. It took a while, but when they discovered that Habibe was pregnant again they were delighted. They went to their first consultation with the midwife together. To everyone's surprise, the ultrasound revealed not one, but two beating hearts. Habibe was pregnant with twins! That was a big shock. A twin pregnancy is a lot harder, after all.

Everything was fine until around six months, but then Habibe's bump got so big that it started causing problems. They also discovered that one of the babies was a bit smaller than the other. Habibe was advised to get more rest. That was easier said than done with a three-year-old at home and a six-year-old who needed to be taken to and from school. Fortunately, Habibe's mother was able to help out.

It was a tense time. Habibe didn't feel great and was worried about her twins. She was sleeping poorly too, which wasn't helped by the fact she couldn't find a comfortable position to lie in with her huge bump. The two babies in her belly were also active and their kicking kept her up at night. The lower baby was lying with its head down, but the smaller one on top kept changing position.

Five weeks before the due date, Habibe suddenly went into labor. The contractions were intense but didn't cause her to become very dilated. The staff in the hospital decided to give her medication via an IV drip to stimulate her contractions. It soon became clear that the smallest child wasn't responding very well, and a cesarean section was suggested.

Habibe was shocked. She was scared to death of having a cesarean section. Fortunately, her husband was with her and that helped calm her down a bit. There were lots of people in the operating room, but everyone was very kind to Habibe. There was a woman who stayed beside her head the whole time and explained everything to her. First, a spinal block was given, which Habibe found unpleasant. She didn't like having to bend forwards with her big belly or the fact that everything was happening literally behind her back. But it was surprisingly quick.

The operation had only just started when Habibe started feeling really nauseous. It got so bad she threw up, but she felt a bit better again after that. Then came the big moment when the two babies were born. Habibe talks about it proudly. "They were lifted out of my belly," she beams. "I could see it clearly. It was two girls. One was quite a bit smaller than the other. The smallest one went in the incubator, but the other one was immediately put on my chest. They put her under the tight purple cloth they'd put on me before they performed the cesarean section." She sighs and says, "It still seems like a miracle that my children were born like this."

Cesarean section

A cesarean section (or C-section) is performed in around 30 percent of births in the US, UK, and Canada. In some countries the rate is much lower, like the Netherlands, where it's only 17 percent. But it's much higher in some countries, like Brazil, where 55 percent of children are born via cesarean section. In many countries, the cesarean section is regarded as a luxury option that wealthy women choose to avoid having to go through the pain of labor or vaginal stretching. In the late 1990s, Corien lived in a town in the north Bolivian rainforest with her first child. When it came up in conversation that she had given birth vaginally, she was asked on more than one occasion, "But why? Didn't your husband let you have a cesarean section?"

There are various reasons why a woman may have a cesarean section. Sometimes the woman may have a planned cesarean section, such as if the child is in the transverse lie position, if the woman has a multiple pregnancy with an unfavorable position, in the event of an infection of the birth canal with a virus like HIV, or if there have been serious problems in a previous pregnancy. Emergency cesarean sections may also be performed, for example if the baby is in imminent danger due to placental abruption or if assisted birth has been unsuccessful. If there is an acute risk to life, a cesarean section can be carried out within a few minutes. In these cases, there is no time for an epidural, so a general anesthetic is given.

The term "cesarean section," or *sectio caesarea* in Latin, refers to the story that Julius Caesar was born in this way 100 years BCE. This story is based on an incorrect interpretation of ancient texts and is almost certainly untrue. The earliest report of a cesarean section being performed and the mother surviving the intervention dates back to the fourteenth century, but is probably also based on a myth. The first person to perform a successful cesarean section was, in fact, probably a Swiss veterinarian around 1500. His wife labored for many hours before he delivered their baby via cesarean section. There was no anesthetic. The woman managed to survive the operation, recovered well, and went on to have five more children.

The number of cesarean sections performed remained very low in the centuries after this initial success. It was only in the twentieth century, when anesthesia became available and surgical techniques improved, that the intervention started to be performed more often. Previously, the surgeon opened up the abdomen by making a substantial vertical incision. This technique is no longer carried out today. Instead, the "bikini cut" is the norm: a small horizontal incision in the lower abdomen measuring no more than six inches. It only takes a few minutes to make the incision. The challenge is opening the uterus at exactly the right place: where the cervix meets the uterus. The gynecologist then manually feels for the baby's head and the baby is carefully lifted out of the abdomen.

As the anesthesia for a cesarean section is usually administered via an epidural or spinal block, the mother can consciously experience the birth alongside her partner. The partner stands with the mother and a screen is put up so they don't have to see her belly being opened up. After birth, the child is put on the mother's chest where possible and she is sometimes even able to pick up the baby herself.

After a cesarean section the incision needs to be closed, taking around twenty minutes. The scar in the uterus doesn't always heal properly. In around 60 percent of cases a small niche, or pouch, is formed where the layers of muscle haven't healed properly. This is often asymptomatic, but women may notice it when their periods return. Mucous and menstrual blood may get left behind in the pouch, which can lead to spotting after menstruation. This niche can sometimes be painful.

Niches are a relatively new phenomenon. They are a lot more common these days because many more cesarean sections are being performed now than twenty or so years ago. If the blood loss from the niche significantly affects the woman, the raised edge of the niche in the uterine wall can be smoothed out with the help of a hysteroscope. It's not yet clear whether the uterine wall is weaker where the niche is and whether it forms a risk for any future pregnancies.

The question as to whether it's possible to have a vaginal birth after a cesarean section is not always easy to answer. The risks need to be identified first. If the baby is not too large and has dropped down sufficiently in the pelvis, a vaginal birth is usually possible. There is, however, a very slim chance that the old scar will open up again due to the force it is under during delivery. In that case, an emergency operation will be necessary. That's why a hospital birth is always advised after a previous cesarean section.

Facts about giving birth

- The record for the woman who gave birth to the most children goes to Valentina Vassilyev from Russia. Between 1725 and 1765 she is said to have given birth to sixty-nine children, including sixteen pairs of twins, seven sets of triplets, and four sets of quadruplets. All but two children survived infancy.

- In newborn babies, the head makes up a quarter of the body. In an adult, the head only accounts for an eighth of the body.

- Only 4 to 5 percent of women give birth on their due date.

- The heaviest babies are born in the month of May. On average, they weigh close to half a pound more than babies born in other months.

- A baby is born somewhere in the world every four seconds.

- A baby is born with 10,000 taste buds. That's forty times as many as an adult, who has 250.

- Newborn babies can breathe and swallow at the same time.

Childbirth and mortality

Modern medical techniques, such as continuous monitoring of the baby's heart rate and the option to have an assisted birth or cesarean section if necessary, have increased the survival rates of mother and child. On top of this, there have been other changes such as improved hygiene, the introduction of imaging techniques, anesthetics, and antibiotics. All of this has meant that death relating to childbirth has decreased significantly over the past century. In the United Kingdom today, according to the University of Oxford, the maternal death rate is 13.41 women per 100,000 births; according to the CDC, it's 32.9 women per 100,000 births in the US. By way of comparison: this figure was 69 mothers per 100,000 births in the 1950s in the UK.

The mortality rate of babies before or during delivery has also decreased over the same period. In 2021, the chance that a baby in the US wouldn't survive pregnancy and birth was 5.4 per 1,000 babies, according to the CDC. In 1935, that figure was 60 deaths per 1,000 births. While things have improved by a large measure, this doesn't take away from the fact that stillbirth and maternal death still occur and cause great grief and sorrow.

7

The Non-Fertile Years

*On the half of life in which
the uterus cannot give birth*

THE UTERUS CAN BLEED and cramp, grow, blossom, and birth. Then it can shrink again, bleed again, perhaps bleed even more heavily. Then stretch for a second time, get large and round again, birth again. And if it has to, it can keep doing this time and time again. During the fertile years, the uterus is a dynamic organ that often makes its presence felt.

The only thing is: those fertile years currently make up at most half of the average lifetime of a person with a uterus. How does the uterus function in the non-fertile years? What signs do we see of a uterus that is not yet or no longer in operation?

The very beginning

Early in the embryonic phase, at six weeks, a girl's uterus, fallopian tubes, and ovaries develop. A female fetus at twenty weeks has between five and seven million egg cells in her ovaries. That's the most she will ever have. Further eggs are not produced during a lifetime and millions are already lost before birth.

When a baby girl is born, everything is present in miniature: the uterus measures no more than half or three-quarters of an inch. The fallopian tubes are barely visible, and the ovaries are tiny, but contain around two million eggs. In a newborn baby, the wall of the uterine muscle is still small and thin, but the uterine cavity already exists. And a thin layer of mucous membrane is already present on the inside of the miniature uterus.

Occasionally a newborn baby girl bleeds from her vagina, but usually no more than a few drops. This happens because the baby was exposed to her mother's pregnancy hormones via the placenta when she was in the uterus. At birth, the exposure to pregnancy hormones stops very abruptly. The endometrium in the child's uterus sometimes responds to this by bleeding a little.

The uterus doesn't generally make itself known during the first ten years of life. It grows along with the child, who develops from a baby measuring around twenty inches in length to a five-foot-tall girl. On average, a ten-year-old girl weighs around ten times as much as she did when she was born. During the first ten years of life, the uterus gradually grows to over an inch in diameter and the ovaries grow too. The number of egg cells continually decreases, and by the start of puberty, only 300,000 to 500,000 remain.

It's possible to see a ten-year-old girl's uterus via ultrasound, albeit with some difficulty. This may be necessary if a girl has serious abdominal symptoms. As transvaginal ultrasounds are not advised at that age, the ultrasound probe is moved over the abdomen. It's not easy to see the relatively small uterus. It can be helpful if the bladder is full, as a full bladder helps transmit the ultrasound waves.

at birth

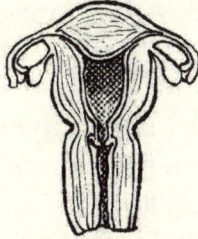

puberty

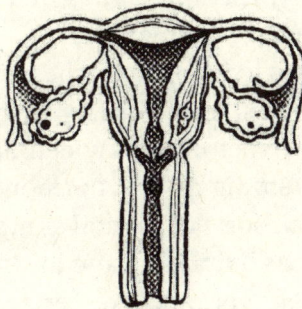

adult

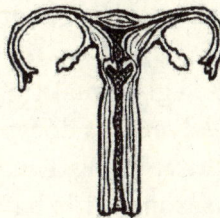

postmenopausal

*The uterus throughout a lifetime. Growth and shrinkage
of the uterus over the years.*

Puberty

When, under the influence of hormones, the uterus starts preparing itself for its real work, puberty begins. This usually happens at the age of ten or eleven, when the ovaries start producing the female hormone estrogen. They receive the signal to do so from two structures in the brain, the pituitary gland and the hypothalamus.

It's not through her uterus that a girl knows she's started puberty, but through the fact that her breasts start to develop and she begins to see underarm and pubic hair. But her uterus, hidden deep inside the belly, has also started a growth spurt. At the onset of puberty, it's the size of a strawberry. By the time a girl's first period arrives, it has grown to the size of a small pear.

A girl's first period, or menarche, can be expected a year or two after the start of puberty. In the meantime, her breasts develop further, her hips widen, and pubic hair and armpit hair thicken. Significant hormonal changes in the body can cause all sorts of mental symptoms to arise in this period, such as irritability and low mood. Puberty is usually a turbulent period characterized by uncertainty and extreme highs and lows. Sometimes the skin also responds to the hormonal changes and acne develops on the face and back.

In the US and UK, a girl usually has her first period at around the age of twelve. This doesn't necessarily mean that the girl is already ovulating. There are big differences between individuals: some girls have already ovulated before their first period, but for around half of girls it can take a couple of months or longer before they ovulate for the first time. In that case, the body only makes estrogen in the first few months and not progesterone. These girls are therefore menstruating, but are not yet fertile.

Girls who don't initially ovulate may experience heavy, painful, and prolonged bleeding. This is because progesterone is not yet being produced. This is called menorrhagia around menarche. It is, in itself, not dangerous, but girls usually find it bothersome. A girl who hasn't yet menstruated doesn't know what's happening to her if her period suddenly starts and it's really heavy and won't stop. What's more, the heavy blood loss is often associated with menstrual cramps and can lead to anemia.

In these cases, GPs usually prescribe the contraceptive pill. This doesn't halt the process of maturation the way some people think. Hormonal development continues but its symptoms are usually masked by the use of the pill. The pill puts a stop to the heavy bleeding and pain, which is often a huge relief. It's important that in these cases, GPs take the time to properly explain what's going on and what the treatment options are. It's not usually necessary for a girl to keep taking the pill for many years if she doesn't want to. After a year and a half she can stop taking it to see if her menstrual cycle has found calmer, more balanced waters.

GPs are relatively quick to prescribe a low-dose contraceptive pill to girls with menstrual problems, and that often works well. But discussions have been arising in recent years about the use of the contraceptive pill in young women. That's partly due to the publication of books about the menstrual cycle, such as Maisie Hill's *Period Power*. Some women say that they regret blindly taking the pill during puberty. Their GP had prescribed it without providing any explanation. They simply started taking it when they actually had no idea how their cycle worked and the effect that the pill had on it.

Now, thanks to recent publications and podcasts, young women are starting to think more about menstruation and the importance of the cycle, and some are starting to take a fresh look at the time they started menstruating. In an episode of the podcast *Damn, Honey*, Marie Lotte Hagen says, "I'm sad for teenage me that I didn't have this knowledge then. I wish I'd known what I do now, back then. That you can simply talk about it and say: it's just started and it's really intense and that's why I'm feeling like this."

The depleting stock of eggs

After a girl's first period, she can expect to be fertile for around forty years. During that time, a woman will have a menstrual cycle every month, any pregnancies aside. Each month, dozens of egg cells in her ovaries prepare for ovulation. Usually only one egg cell becomes dominant and is released. The rest are absorbed into the body.

The stock of eggs in the ovaries therefore depletes over time. When a woman is forty, only 5 percent of her original eggs are left. The cohorts of eggs that prepare for ovulation get smaller and the ovaries become less sensitive to the hormones released from the pituitary gland. This causes the production of estrogen to start to decrease and menopause to draw closer.

The very start of the menopausal transition, or perimenopause, is usually difficult to establish. From the age of thirty-five, a person's cycle starts showing signs of ageing. Many women experience heavier periods at this time and their cycle becomes shorter. These are the first signs that change is afoot.

It's impossible to say how long this phase of heavy, frequent periods will go on for. Sometimes it can last a few years. Then the cycle gradually starts to become less reliable. A period might arrive much later than expected or a woman might not have a period for a long time before it suddenly returns with ferocity. Perimenopause is characterized by its lack of regularity. Women can become overwhelmed by heavy periods that occur suddenly and are accompanied with severe pain and discomfort.

In the meantime, fertility continues to decline and the risk of miscarriage increases. Up to the age of thirty-five, the risk of miscarriage is 15 percent and after the age of forty it increases to 35 percent. The chance of having a child with a congenital disorder also increases significantly with the mother's age, and the father's age too. This reduced fertility is because the stock of egg cells is depleting and the eggs that are left are lower quality. In addition, cell division errors are more likely to occur after the age of forty.

Denise (47) *on perimenopause*

"My whole life's been turned upside down. It's awful," Denise explains. "I'm starting to realize that it's actually been going on for a while. But I hadn't thought about perimenopause until now. I was surely too young for all that."

Denise is the manager of a cleaning company. She's responsible for around sixty employees, who clean various buildings. The staffing, the distribution, and the teams—it's all Denise's responsibility. It's a difficult job that requires flexibility and creativity. Schedules need to be constantly adapted and rejigged due to illness, absences, or different client requirements. Denise is good at her job. She gets a kick out of being able to come up with the right solution every time.

Last year, something changed. Denise is struggling more and more with her work. It requires more effort and she often feels ill-inclined to completely reconfigure a complicated schedule. What's happened to her trademark creativity? It's getting her down. Her home life with two adolescents isn't as smooth as it used to be either. She doesn't have much patience. She gets really exasperated with the pair of them sitting around all day, but they probably get exasperated with their nagging mother too.

Over the past few years, her periods had been getting heavier and heavier. She'd noticed it, but hadn't really paid it a second thought. Now, suddenly, nothing has happened in the three months since her last period. She sleeps poorly and often wakes up in the middle of the night soaked in sweat.

Denise suddenly realizes this might be perimenopause. Could changing hormone levels be why she's been feeling so miserable and unbalanced for a while now? She'd really like to see if there's something that might help her with her symptoms. Her friend told her that she'd started taking the contraceptive pill again. Could that be something for her too? She doesn't smoke, doesn't take any other medications, and was relatively happy with the pill when she took it between the ages of seventeen and twenty-eight. That was before she had children. Once her family was complete, her husband was sterilized so she no longer had a need for it.

Denise goes to her GP and is prescribed Zoely, a contraceptive pill that contains a natural estrogen. Most of the other pills contain a synthetic estrogen. It doesn't take long for Denise to notice the difference. She feels much better and has started enjoying her hectic job again.

Menopause

Then comes the day when the stock of egg cells has become almost entirely depleted, and the ones left are of a poor quality. That's when the ovaries definitively stop producing estrogen. Menopause is a reality. The fertile years have come to an end.

The exact moment of menopause can only be determined retrospectively. Once menstruation has been absent for a year, you can say that menopause has happened. One study reported that the average age of menopause is 54 in Europe and 51.4 in North America, but there is lots of variation; some people become menopausal at the age of forty, whereas others continue menstruating until well over the age of fifty. For one in a hundred women, periods stop before the age of forty; for one in a thousand, they stop before the age of thirty. The age of menopause appears to have a genetic factor. A mother who goes through menopause late is more likely to have a daughter who will also go through menopause late. Women who smoke go through menopause an average of two years earlier than those who don't.

Strictly speaking, menopause is simply the time when fertility has ended. Once the cycle has definitively stopped, the body enters a phase, so-called postmenopause, when it has to find a new balance. Perimenopause encompasses the whole period from the first signs the cycle is starting to age up to and including the moment the body's equilibrium is restored. It's a part of life that can take around ten years, or longer.

Role of the uterus

Menopause renders the uterus inoperative after about forty years of service. The organ doesn't play a major role during perimenopause. The end of fertility isn't caused by the ageing of the uterus, but by the supply of eggs in the ovaries being used up. Without mature egg cells and without the influence of hormones, the uterus can no longer perform its job. It becomes smaller and a period of rest ensues. The endometrium becomes thin. On an ultrasound scan, the uterus appears as a gray oval without a clear cavity. There's a thin white stripe measuring less than a fifth of an inch at the point where the layers of endometrium touch.

However, it's possible to get the postmenopausal uterus working again, if desired. If stimulated by external hormones, an old uterus can often carry a pregnancy with the help of IVF treatment. In Italy, for example, where IVF treatment doesn't have an age limit, there are various examples of women who've had a successful pregnancy well after the age of fifty. And a seventy-year-old Ugandan woman named Safina Namukwaya gave birth to healthy twins following IVF in 2023.

Transplantation of a postmenopausal uterus is sometimes also possible. If, for example, a daughter doesn't have a uterus due to an illness or abnormality, her own mother's uterus could be transplanted, even if the mother has already gone through menopause. In a young body, the transplanted uterus will simply resume its tasks under the influence of hormones. If all goes well, the daughter can then become pregnant with the help of the uterus from which she herself originated.

Marlies (64) *on hot flashes and other discomforts*

I was fifty-five when I went through menopause. Until then I hadn't had any problems. Funnily enough, I don't remember my menstrual pattern changing much. Maybe my periods suddenly stopped, or maybe there were longer gaps between them—I simply can't remember.

One thing I do remember is the hot flashes. All of a sudden, I'd be boiling hot, and then I'd be cold again. Obviously other people didn't notice them, as I sometimes had them during consultation hours and nobody ever commented on them. But perhaps they were just being polite.

I also remember the night sweats. I'd be so hot that I'd push my feet and legs out from under the duvet, only to be freezing cold again a bit later. At that time I also noticed that I wasn't able to sleep during my night shifts as easily as I had done in the past after I'd been called out to assist a birth or for a consultation. That was a real shame. Night shifts became much more difficult as a result.

Fortunately, I was able to carry on working throughout perimenopause and I wasn't too affected by mood changes. I later sometimes asked myself why I didn't take hormone supplements. That would have been handy for all the wrinkles and skin changes too. I wish I'd spent several years using hormone creams you can just rub into your skin.

I'm aware how fortunate I was. Menopause didn't completely derail my life the way I often see in my patients. If there's one thing I've learned in all these years as a gynecologist, it's that you're not the one calling the shots. You simply have to wait and see what menopause is like for you. You can try your hardest to make the most out of it, but it can still be a massive disappointment.

Women need to be kinder to themselves. Menopause is a difficult time in which you can pamper yourself or allow yourself to be pampered a bit more. You really don't need to pretend there's nothing going on.

Ligaments

Although the uterus itself isn't usually affected by ageing, the ligaments that support it and hold it in place may become looser over time. This usually only happens after menopause, when there's no longer much estrogen in the body, which causes the tissues to lose their elasticity. This can lead to uterine prolapse, when the uterus drops down into the vagina. Sometimes the vaginal wall also weakens, which can cause the bladder and intestines to prolapse.

There are several ways to treat prolapse. A prolapse of the vaginal wall can be supported by a pessary—a prosthetic, often a ring, that can be inserted into the vagina to hold organs in position. Surgery to support the vaginal wall is also possible. The excess tissue of the vaginal wall can be reduced. If the uterus has only prolapsed because the ligaments have become loose, the ligaments can be shortened. After this type of surgery, everything may be fine for a number of years, but if the less elastic vaginal wall starts to stretch again over time another prolapse may occur.

In the battle against prolapse, transvaginal mesh implants introduced in 2005 seemed like just the thing. The mesh was surgically inserted under the vaginal wall below the bladder or above the rectum. The mesh laid a base to reinforce the vaginal wall and prevent the front and back walls of the vagina from prolapsing. However, in recent years this procedure was banned in the US, UK, and Canada because health authorities concluded that the risks—like bleeding and severe pain—outweighed the potential benefits.

Symptoms

Beyond cycle changes and reduced fertility, for many women perimenopause is characterized by physical and mental symptoms. More than three quarters of women struggle with at least one health issue during perimenopause and 40 percent report three or more issues.

The problems are caused by hormonal disruption, especially the shortage of estrogen in the body. The drop in estrogen impacts not only the cycle, but also things like hair growth, skin thickness and elasticity, bone density, vaginal wall thickness and moistness, blood vessel condition, brain functioning, and bladder capacity.

The sudden drop in estrogen levels at the start of menopause means that the ageing process is different in women than men. Male fertility doesn't end abruptly, although the production of the hormone testosterone—the male counterpart to estrogen—gradually decreases over time. This decrease can also cause symptoms in men, but these are generally less sudden and less intense than the symptoms women experience.

The Flemish gynecologist Herman Depypere writes in his book *Menopause: All Questions Answered*, "During menopause, there is a significant difference between man and woman and nature 'discriminates' against the woman by fully eliminating the female hormone." According to Depypere, women are "severely disadvantaged by nature" and therefore worse off than men in this regard. But he hastens to add: there's something that can be done about many of the problems that women experience as a result of a shortage of estrogen.

Not all women experience symptoms to the same extent. Women who take the contraceptive pill don't tend to notice the effects of menopause much, as the pill masks the process. It's only when they stop taking the pill that they discover that their periods have stopped. Women with an IUD don't usually notice that their periods have definitively stopped either, as they hardly bleed or no longer bleed anyway due to the device. However, they may experience symptoms of menopause. As an IUD only contains progesterone, someone using one can still be affected by the shortage of estrogen experienced during menopause.

Symptoms of perimenopause:

- Hot flashes

- Night sweats

- Sleep problems and fatigue

- Headaches and migraines

- Problems with concentration and memory

- Mood changes and irritability

- Depressive symptoms

- Anxiety and panic attacks

- A dry, burning sensation in the vagina

- Pain during sex

- Reduced libido

- Stiff joints and muscles

- Osteoporosis

- High blood pressure

- Heart palpitations

- Thinner and drier skin, more wrinkles

- Brittle nails

- Thinning hair

- Fat accumulation around the belly

- Feeling frenzied, restless

- Tightness in the chest, shortness of breath
- Weight gain
- Bladder problems, such as frequent urge to pee, urinary incontinence
- Unwanted hair growth on the face

The list of issues that can arise during perimenopause is long and varied. The sudden hot flashes are one of the best-known symptoms. They occur because hormonal fluctuations affect the regulation of body temperature. The body produces adrenaline during a hot flash, which results in raised blood pressure, heart palpitations, blood vessel dilation, and heavy sweating. A hot flash can last between one and three minutes. Night sweats are also a result of problems with body temperature regulation. Sudden, intense sweating episodes in the middle of the night are a common occurrence during perimenopause.

As we're writing about perimenopause, we clearly recall our own experiences of hot flashes. How awful it was when we were at work and we would suddenly break out in a sweat all over. The interrupted nights when we would wake up with a start, soaking wet, throw the covers off, only to be shivering again with the cold fifteen minutes later. Virtually everyone experiences this.

Selin (54) *on her hysterectomy and menopause*

Selin had her uterus removed at the age of forty-six. She'd suffered from large fibroids that caused extremely heavy blood loss. She'd really wanted to keep her uterus, but the fibroids had extended as far as her navel. A hysterectomy was the only option.

Fortunately, it was possible for the intervention to be carried out via laparoscopy, so she doesn't have a large abdominal scar. She just has very small marks in three places and in her navel. As her uterus was removed, Selin hasn't had a period for many years. It's been a great relief for her not to have to contend with those heavy periods anymore. She's found joy in life again and has more energy.

However, things recently took a turn for the worse. She sleeps poorly, sometimes tossing and turning all night long. She feels hot and throws the duvet off, only to then feel shivery and pull the covers back over her again. The worst part is the night sweats. She sometimes sweats so much that her whole bed is soaked, and she has to change her sheets in the middle of the night. Her neighbor told her that there are hormones you can take to help. That seemed like a great idea, but Selin's husband insisted she shouldn't take them.

Finally, Selin decided to go to the doctor after all. Her doctor explained that she only needed the hormone estrogen as she didn't have a uterus anymore. Estrogen reduces symptoms of menopause, but also stimulates the endometrium, which isn't a problem for Selin. Estrogen is available as a gel or patch. Applying estrogen to the skin once or twice a day is said to help reduce symptoms.

After three months, Selin called the gynecologist and sang the praises of the estrogen treatment. Her sleep is back to normal again and she feels just as energetic as she used to. She wanted to know how long she could continue using it for. The doctor told her she can carry on using it for at least five years or so, then gradually reduce the dose.

Treatment

Symptoms of perimenopause vary and it's not easy to predict which symptoms will occur and to what extent. The differences between women are huge. One woman may have serious health issues while another barely notices anything. For women with symptoms that aren't too severe, proper information and lifestyle advice related to healthy eating, rest, relaxation, and exercise can make a big difference. In addition, simple, practical solutions can be considered, such as flexible working hours for women who are going through a period of poor sleep or the opportunity to take a break from the working day in the event of severe hot flashes.

However, if the symptoms are considerably impacting a woman's day-to-day life, other solutions may be necessary. That's the case for more than 25 percent of women going through perimenopause. There's a relatively large number of women, for example, who experience faster bone deterioration due to a lack of estrogen, which can lead to osteoporosis. According to the CDC, 27.1 percent of women over age sixty-five have osteoporosis. In the US and Canada, more women die of osteoporosis than breast cancer.

For women with serious symptoms of perimenopause, menopausal hormone therapy (sometimes called hormone replacement therapy) can be a good option. The idea behind it is logical: the problems are caused by hormonal disruption and a shortage of estrogen. Could that shortage not be treated by replacing hormones in a subtle manner?

In the UK, only around 10 percent of women with serious symptoms of menopause receive menopausal hormone therapy (MHT). In the United States, the rates of MHT have dropped off dramatically after a study linked MHT to breast cancer and cardiac risks, and today, only around 4 percent of women take the treatment, down from a high of 22 percent.

While the approach towards MHT can be cautious, specific groups may be advised to take hormones for a while. This includes women who become menopausal before the age of forty. This is called primary ovarian insufficiency, or premature ovarian failure. As these women experience a shortage of estrogen and progesterone at a much younger age than usual, their risk of osteoporosis is much higher.

The use of a hormone preparation can slow down the process of osteoporosis in these women. This could be a contraceptive pill that contains estrogen and progesterone, or MHT tablets that contain a low dosage of both hormones. If a woman has had a hysterectomy, she will not need to take progesterone, but otherwise it is not a good idea to administer estrogen on its own as this causes the endometrium to be constantly stimulated. It can then form a thick layer that bleeds erratically. Furthermore, long-term exposure to just estrogen can, in some instances, cause endometrial malignancies.

Though the approach to MHT is often conservative, today, more and more people are calling for hormone supplements to be considered in certain circumstances. The increased risk of breast cancer appears minimal. MHT also has many advantages: it is effective for symptoms of menopause such as hot flashes and mood changes, decreases the risk of osteoporosis, and may decrease the risk of cardiovascular disease. Research has even shown that after ten years of hormone therapy, women live on average two years longer than women who didn't have hormone therapy.

New hormone medication is now available that has fewer side effects. This medication can be applied to the skin with a patch, which means that the hormones don't have to be administered in such high dosages. The skin absorbs the drugs and transports the estrogen to the blood vessels to be circulated through the body. Progesterone often has to be taken in pill form, while progestin is administered via an IUD.

Taboo

Women between forty and sixty generally lead an active life and have all sorts of roles and responsibilities in the fields of work, care, and volunteering. As women today have children relatively late, they often have adolescents still living at home with them at around the age of fifty. A majority from this age group are working, and often find themselves at the height of their career.

And then perimenopause arrives and around 80 percent of women in this age group experience health issues, sometimes to such an extent that their day-to-day functioning is impaired. In many ways, perimenopause seems to mirror puberty, which marked the start of a woman's reproductive years. Both before the first period as well as around the time of the last period, women experience a turbulent time. In both situations, hormones are disrupted, which can cause all sorts of symptoms that may also affect the brain.

However, we seem more willing to accept that girls in puberty might be a bit unbalanced or experience other symptoms compared to women in perimenopause. Puberty is seen as a turbulent but promising time. The start of fertility is something to look forward to. Menopause, on the other hand, is generally seen as anything but promising. The end of fertility is often seen as doom and gloom: wrinkles, brittle bones, and hot flashes in combination with feeling like you're no longer attractive and don't really count anymore.

Research also seems to pay more attention to hormonal disruption during puberty—which also, incidentally, affects boys—than to disruption during perimenopause, which is in many ways comparable. The "adolescent brain" has become an established concept over the past years, in part because of the publication of successful books about brain development in adolescents.

What's less well known is that a "menopause brain" probably also exists. The presenter and journalist Elles de Bruin drew attention to this in her podcast *Hot Flashes* in 2020. She pointed out that doctors and healthcare providers often fail to acknowledge that mental and emotional symptoms such as mood changes are associated with perimenopause. In many cases, this link is not even considered. As a result, women are often wrongly diagnosed with burnout or depression. Medication is prescribed and the women end up off work sick. It should be standard for doctors and healthcare professionals to ask women between the ages of forty and sixty about symptoms of perimenopause, but in many cases, they fail to do so.

How can there be so much ignorance about a life experience associated with such serious health issues? Why isn't more research into menopause being carried out? There still seems to be a taboo surrounding this phase of life, which every woman goes through at some point. People don't speak up about the topic due to anxiety or embarrassment. And if people do talk about it, they generally do so in the same way people talk about menstrual issues: "It's all part and parcel of it, you have to get through it, don't pay too much attention to it and, most of all, don't talk too much about it."

Perimenopause isn't seen as a fun topic in a world that revolves around being young and sexy. As a result, there's a serious lack of knowledge about what perimenopause entails. Lots of people don't know the specifics, but they're not interested in finding out about them either. All too often, women seem unprepared. Many women say perimenopause came as a complete surprise to them. They didn't recognize their symptoms because they hardly knew anything about perimenopause, so they had no idea what was happening to them.

Over the last few years, things have started to change. The topic of menopause is gaining interest, which can be seen in books, documentaries, websites, theater shows, webinars, and podcasts. Menopause clinics and special consulting sessions are emerging, and there are more health professionals who specifically deal with problems relating to menopause. And in the US, a bill introduced in Congress in 2022 called for $100 million for menopause research; in the UK, a law was proposed (but rejected) to make menopause a "protected characteristic" under the Equality Act.

This attention to the topic is sorely needed because the impact of menopause is much greater than most people think. In Canada, 32 percent of working women say their menopause symptoms negatively impacted their performance at work; in the UK, a survey showed that 14 percent of women are considering quitting due to their symptoms. Each year, the costs associated with this are massive. The employment rate of women in perimenopause also drops, although it's hard to establish whether perimenopausal symptoms are the main cause of this. Occupational doctor Mariëlle van Aalst, who sees lots of women with perimenopausal symptoms in her consulting room, is clear on this: "Menopause is a real career killer."

New phenomenon

It's important to recognize that menopause is a relatively new phenomenon. In the past, most women simply didn't live long enough to reach menopause. You could even say that a substantial period of life after the end of fertility wasn't intended from an evolutionary point of view, as gynecologist Herman Depypere claims in his book *Menopause: All Questions Answered*. Menopause is rarely seen in the animal kingdom. Most animals die not long after their fertility comes to an end. Only killer whales, short-finned pilot whales, rhesus monkeys, macaques, and chimpanzees are known to continue living after menopause. Some insects also appear to continue living after they have reached the end of their fertility.

For many centuries, the vast majority of women died before the end of their reproductive years, sometimes a long time before. Menopausal women were an exception, and that didn't always have positive repercussions. There was a strikingly high number of older women who no longer menstruated among the "witches" persecuted in the Middle Ages. Old age and infertility were perceived as a threat. Most of the persecuted women were not only old, but also unmarried and lived in poverty.

Menopause only became more frequent when life expectancy started to increase. At the beginning of the twentieth century, women didn't generally live to much older than fifty. A century later, women in the developed world have gained an average of more than thirty years.

Although life expectancy has increased quickly over the course of a century, the age at which menopause starts hasn't changed. Why not? In 2013, a group of (male) Canadian researchers came up with a controversial explanation for this. They made a computer simulation of human reproductive characteristics and concluded that fertility is no longer useful for older women as men prefer younger women. According to the researchers, the reverse would also apply: if women were to favor younger men, older men would eventually also lose their fertility.

An alternative, popular theory is the Grandmother Hypothesis, which was first introduced by ecologist George C. Williams in 1957 but was popularized by the anthropologist Kristen Hawkes beginning in the 1980s. She found that the foraging of older women in hunter-gatherer communities in Tanzania allowed younger women to have more babies—which would give societies where women live longer an evolutionary advantage.

Whatever the reason, the fact is that most women still have around a third of their lives left to live when they become menopausal. All the more reason to focus on the question: How can we make life with a uterus that is no longer operational as healthy and enjoyable as possible? A lot more research into this is required. Above all, we need to do away with the taboo surrounding this phase of life and stop referring to it in terms of old, past it, and uninteresting.

Glossary

Abortus provocatus

Termination of pregnancy.

Adenomyosis

In this condition, the lining of the uterus starts growing into the muscle in the wall of the uterus.

Amenorrhea

The absence of or cessation of the menstrual period.

Breech delivery

Delivery of a baby whereby its lower body comes first, as opposed to its head.

Carcinoma of the cervix

Cancer of the cervix. This is often caused by the human papillomavirus (HPV).

Carcinoma of the ovary

Ovarian cancer.

Cervical screening

Previously called the "Pap smear." A vaginal examination whereby mucous membrane cells are removed from the cervix.

Cervix

The lower part of the uterus, which has an opening into the uterus called the internal os and an opening into the vagina called the external os.

Cesarean section

Surgery to deliver a baby via incisions in the lower abdomen and uterus.

Coitus interruptus

Literally: interrupted sexual intercourse. During penetrative sexual intercourse, the penis is removed from the vagina prior to ejaculation, with the intention of preventing pregnancy. Also known as the withdrawal method.

Conception

Fertilization.

Contraception

Means or method used to prevent pregnancy.

Contraceptive implant

A contraceptive in the form of a plastic rod placed under the skin in the upper arm. It releases hormones into the bloodstream.

CTG

Cardiotocography. A technique used to monitor the fetal heartbeat during labor.

Cyst

A sac in the body that usually contains fluid.

Dermoid cyst

A cyst that usually develops before birth and can contain sebum, hair, and other types of tissue, such as nails and teeth. Generally benign.

Diaphragm

Or: cap. A method of contraception in the form of a ring that's inserted vaginally to block the cervix so that sperm cells cannot enter the uterus. Should be used together with a spermicide.

Dilation

Process during labor whereby the cervix opens under the influence of dilation contractions.

Early medical abortion

Termination of pregnancy that takes place within the first ten weeks of the pregnancy.

Effacement

The process prior to delivery during which the cervix softens and then disappears completely so that the uterus and vagina can together form the birth canal.

Embolization

A medical technique carried out by a radiologist using fluoroscopy to block off blood vessels that lead to a fibroid, for example.

Emergency contraception

Emergency contraception can prevent pregnancy after unprotected sex. It comes in the form of the morning-after pill or the IUD.

Endometrial ablation

Treatment for very heavy menstrual flow, whereby the endometrial lining of the uterus is removed or destroyed.

Endometrial carcinoma

Cancer of the endometrium on the lining of the uterus. Also known as uterine cancer.

Endometriosis

In this condition, tissue similar to the uterine lining (the endometrium) grows outside the uterus. It is a benign condition that can cause severe pain and heavy bleeding.

Endometrium

The inner lining of the uterus.

Epidural

An injection in the back just outside the sac of fluid around the spinal cord. It anesthetizes the lower body, for example in the event of intense pain during labor.

Estrogen

Female sex hormone produced in the ovaries. It is involved in various bodily processes, such as initiating puberty. It plays a role in the cycle: the estrogen level peaks just before ovulation.

External os

The outermost opening of the cervix to the vagina.

Extirpation of the uterus

Latin term for hysterectomy.

Follicles

Small sacs of fluid in the ovaries that contain a developing egg. ovaries.

FSH

Follicle-stimulating hormone. It is produced in the pituitary gland and stimulates the growth and maturation of the egg cells in the ovaries.

hCG

Human chorionic gonadotropin. Pregnancy hormone.

Hormone replacement therapy

See menopausal hormone therapy.

HPV

Human papillomavirus. A sexually transmitted infection that can cause cervical cancer.

Hypothalamus

Part of the brain where the hormone GnRH is produced, which stimulates the pituitary gland to make and release FSH and LH.

Hystera

Greek word for the uterus.

Hysterectomy

The medical term for removal of the uterus, originating from Greek. The term extirpation of the uterus is also sometimes used.

Internal os

The point where the narrow cervix meets the wider body of the uterus.

IUD

Intrauterine device. This can be either a copper IUD or a hormonal IUD. The copper IUD, or copper coil, is a small T-shaped plastic and copper device that's put into the uterus. The copper stops a fertilized egg from being able to implant itself. The hormonal IUD is a contraceptive that releases hormones into the uterus. Brand names include Mirena and Kyleena.

IVF

In vitro fertilization. A type of fertility treatment whereby fertilization takes place outside the body.

LH

Luteinizing hormone. It is produced in the pituitary gland and stimulates ovulation.

Ligaments

Bands of tissue that hold the uterus in place in the pelvis.

Menarche

The medical term for a girl/young woman's very first period.

Menopausal hormone therapy

Treatment for serious symptoms of menopause with low doses of female sex hormones (estrogen and progesterone).

Menopause

Once the supply of egg cells is used up at around the age of fifty, the ovaries stop producing estrogen and the menstrual cycle stops.

Menorrhagia

Heavy menstrual bleeding.

Menses

The medical term for menstruation.

Menstruation

The monthly shedding of the lining of the uterus. Also known as a period.

Morcellator

A surgical instrument used in the uterus, for example to divide a fibroid into small pieces and to remove them.

MRI

Magnetic resonance imaging. A medical imaging technique that uses strong magnetic fields and radio waves to produce detailed images of the inside of the body.

MRKH syndrome

Mayer-Rokitansky-Küster-Hauser syndrome. A congenital condition whereby the uterus fails to develop, and the vagina is usually very short. The ovaries are present.

Myoma

The medical term for a fibroid. A growth made of muscle tissue that is formed in the uterine muscle.

Osteoporosis

A disease that weakens the bones.

Oxytocin

A hormone that causes the uterus to contract during labor. Also involved in the production of breast milk.

PCOS

Polycystic ovary syndrome. More follicles than normal develop during a cycle, but ovulation is often absent.

Pearl Index

The Pearl Index expresses the reliability of a specific method of contraception as the number of pregnancies per hundred woman-years of use.

Perineum

The area between the vagina and the anus.

Pituitary gland

A gland found at the base of the brain. This is where follicle-stimulating hormone (FSH) and luteinizing hormone (LH) are produced, which stimulate egg maturation and ovulation from the ovary.

PMDD

Premenstrual dysphoric disorder. A serious form of PMS. A very intense response to the hormone fluctuations in the second half of the cycle. Can be associated with serious symptoms such as depression and concentration problems.

PMS

Premenstrual syndrome. An intense response to hormone fluctuations in the week before the period. Can cause low mood, irritability, and headaches.

POF

Premature ovarian failure. Also known as: primary ovarian insufficiency. When the ovaries stop functioning prematurely (before the age of forty). This causes an early menopause.

Polyp

A growth in the endometrium that is usually benign.

Progesterone

Female sex hormone produced in the ovaries in the second half of the cycle. Maintains the endometrium to enable a fertilized egg cell to implant itself.

Prostaglandins

Hormones that help ripen the cervix in labor and produce the hormone oxytocin, which triggers contractions.

Sectio caesarea

Medical term for cesarean section, derived from Latin.

Speculum

A medical tool used for vaginal examination.

Spinal block

A type of anesthesia given in the spinal cord to temporarily anesthetize the lower body, for example for a cesarean section.

Sterilization

A procedure to prevent pregnancy.

Thrombosis

When a blood vessel gets blocked by a blood clot.

Toxic shock syndrome

A rare but dangerous infection that can be caused by a tampon being left in the body for too long.

Ultrasound

An imaging method that uses sound waves to produce images of structures within the body.

Uterine artery

Artery that supplies blood to the uterus.

Vacuum extractor

Or: vacuum pump. A tool used if labor is not progressing.

Vaginal ring

Contraceptive in the form of a plastic ring that is inserted into the vagina and releases hormones.

Vasectomy

Male sterilization.

VEMA

Very early medical abortion. Pregnancy termination at a very early stage, immediately after a positive pregnancy test.

Acknowledgments

WRITING AN EXHAUSTIVE and comprehensive book about an organ as versatile as the uterus is an impossible task. We have attempted to make an interesting selection from the many stories and facts there are to be told about uteruses and fallopian tubes. We realize that important aspects will undoubtedly have been left out and that relevant sources will have escaped our attention.

We are aware that not only women have uteruses. Transgender men and nonbinary people can also have a uterus. However, for the sake of readability we have generally used the word "woman" or "women" to refer to a person or people with a uterus. It was not our intention to exclude anyone by doing so.

As we didn't want the text to be full of notes, we decided to include all the sources we used in the literature list at the end of the book. The list is structured per chapter.

The portraits that feature in the various chapters are based on interviews and patient stories. Most of the names have been changed to maintain confidentiality. Occasionally, we also altered the facts. The interviewees gave permission for their stories to be included. We also added a couple of stories from our own experiences.

Many thanks to the following people for sharing their stories: Chiara, Christine, Denise, Gizem, Habibe, Irena,

Karin, Kimberley, Louise, Roos, Sanne, Selin, Shainy, Sophie, and Wendy.

For her infectious enthusiasm and involvement: Esther Hendriks.

For the beautiful and detailed illustrations: Gijs Klunder.

For reading carefully and critically and for sharing their thoughts: Yolande Appelman, Marguerite Hoogland, Peggy Geomini, Erik Gramberg, Rodie Gramberg, Pieter Zuidema, Lisa Zuidema, and the members of Corien's writing group: Paulien Bakker, Menno Bosma, and Alfred van Cleef.

For the beautiful English translation and the careful editing: Alice Tetley-Paul and Jennifer Croll.

Literature

Foreword

"Canada's fertility rate reaches an all-time low in 2022," Statistics Canada, January 31, 2024, www150.statcan.gc.ca/n1/daily-quotidien/240131/dq240131c-eng.htm.

Ivana Saric, "Births dropped in 2023, ending pandemic baby boom," *Axios*, April 25, 2024, axios.com/2024/04/25/us-births-drop-2023.

Heather Stewart, "Birthrate in UK falls to record low as campaigners say 'procreation a luxury,'" *Guardian*, February 23, 2024, theguardian.com/uk-news/2024/feb/23/birthrate-in-uk-falls-to-record-low-as-campaigners-say-procreation-is-a-luxury.

Chapter 1

Boston Women's Health Book Collective, *Our Bodies, Ourselves: A Book by and for Women* (Simon & Schuster, 1973).

M. Brännström, M. A. Belfort, and J. M. Ayoubi, "Uterus transplantation worldwide: Clinical activities and outcomes," *Current Opinion in Organ Transplantation* 26, no. 6 (December 1, 2021): 616–26, pubmed.ncbi.nlm.nih.gov/34636769.

Canadian Partnership Against Cancer, "Cervical cancer screening in Canada," 2021/2022, partnershipagainstcancer.ca/topics/cervical-cancer-screening-in-canada-2021-2022/programs/guidelines.

L. A. R. Castellón et al., "The history behind successful uterine transplantation in humans," *JBRA Assisted Reproduction* 21, no. 2 (April–June 2017): 126–34, ncbi.nlm.nih.gov/pmc/articles/PMC5473706.

Rixt de Boer, "Hysteria—voor een orgasme ging je naar de dokter" [Hysteria—you went to the doctor for an orgasm], *Geschiedenis Beleven*, October 23, 2012, geschiedenisbeleven.nl/hysteria-voor-een-orgasme-ging-je-naar-de-dokter.

M. K. Herlin, M. B. Petersen, and M. Brännström, "Mayer-Rokitansky-Küster-Hauser syndrome: A comprehensive guide," *Orphanet Journal of Rare Diseases* 15 (2020): 214, ncbi.nlm.nih.gov/pmc/articles/PMC7439721.

L. Johannesson et al., "The first 5 years of uterus transplant in the US," *JAMA Surgery* 157, no. 9 (September 2022): 790–97, ncbi.nlm.nih.gov/pmc/articles/PMC9260640.

G. A. Lindeboom, *Geschiedenis van de medische wetenschap in Nederland* [A history of medical science in the Netherlands] (Fibula–Van Dishoeck, 1973).

National Cancer Institute, "Cervical cancer screening," cancer.gov/types/cervical/screening.

NHS England, "Cervical screening: Programme overview," Gov.UK, April 1, 2015, gov.uk/guidance/cervical-screening-programme-overview.

Pieter Sabel, "Alternatieve Nobelprijs voor ontdekking dat orgasme net zo goed werkt tegen verstopte neus als neusspray" [Alternative Nobel Prize for discovering that orgasm works just as well against nasal congestion as nasal spray], *de Volkskrant*, September 10, 2021, volkskrant.nl/wetenschap/alternatieve-nobelprijs-voor-ontdekking-dat-orgasme-net-zo-goed-werkt-tegen-verstopte-neus-als-neusspray~b62639c9.

Sahlgrenska Academy, "World's first birth after uterus transplantation with robot-assisted surgery alone," University of Gothenburg, May 26, 2023, gu.se/en/news/worlds-first-birth-after-uterus-transplantation-with-robot-assisted-surgery-alone.

Dick Schoot and Wim Daniëls, *De Baarmoeder* [The womb] (Prometheus, 2017).

Leonie van de Graaf, "De baarmoeder, een zegen of een vloek? Studie naar middeleeuwse opvattingen over de baarmoeder" [The womb, a blessing or a curse? A study into medieval attitudes towards the womb] (thesis, Utrecht University, 2013), studenttheses.uu.nl/handle/20.500.12932/16709.

Chapter 2

"About heavy menstrual bleeding," US Centers for Disease Control and Prevention, May 15, 2024, cdc.gov/female-blood-disorders/about/heavy-menstrual-bleeding.html.

Y. Bentov et al., "Ovarian follicular function is not altered by SARS-CoV-2 infection or BNT162b2 mRNA COVID-19 vaccination," *Human Reproduction* 36, no. 9 (August 2021): 2506–13, pubmed.ncbi.nlm.nih.gov/34364311.

Lisa Bouyeure, "Menstruatiebloed als twijfelachtig politiek wapen" [Menstrual blood as a questionable political weapon], *HP/De Tijd*, March 30, 2016, hpdetijd.nl/2016-03-30/menstruatiebloed-als-twijfelachtig-politiek-wapen.

Marie Cardinal, *The Words to Say It* (Van Vactor & Goodheart, 1984).

Caitlin Carlson, "For these women, 'biohacking' their periods gives them power," *Wall Street Journal*, February 29, 2024, wsj.com/style/period-cycle-tracking-tech-apple-health-oura-ring-be0288ee.

Katrina Clarke, "Free the period: Why some women choose to free-bleed," CBC, March 8, 2017, cbc.ca/life/wellness/free-the-period-why-some-women-choose-to-free-bleed-1.4015740.

Coronameldingen [Corona notifications], on: *Lareb.nl*, 24 October 2021.

"Covid-19 vaccinations may have caused thousands of menstrual disorders," *NL Times*, December 22, 2021, nltimes.nl/2021/12/22/covid-19-vaccines-may-caused-thousands-menstrual-disorders.

"Cyclus en hormonen" [Cycles and hormones], *Cycle*, cycle.care/nl.

Elma Drayer, "Ik word zenuwachtig van het biologisch determinisme dat menstruatiegelovigen uitdragen" [The biological determinism put forward by menstruation worshippers makes me nervous], *de Volkskrant*, July 29, 2021, volkskrant.nl/columns-opinie/ik-word-zenuwachtig-van-het-biologisch-determinisme-dat-menstruatiegelovigen-uitdragen~bd880406.

Marie Lotte Hagen and Nydia van Voorthuizen, *Damn, Honey*, podcast, episode 34 on the menstrual cycle, January 2020.

Hevigbloedverlies [Heavy periods], hevigbloedverlies.nl.

Maisie Hill, *Period Power: Harness Your Hormones and Get Your Cycle Working for You* (Green Tree, 2019).

Katie Hoare, "Data reveals worrying knowledge gap for men in menstruation education," *Happiful*, July 12, 2021, happiful.com/data-reveals-worrying-knowledge-gap-for-men-in-menstruation-education.

Karen Houppert, *The Curse: Confronting the Last Unmentionable Taboo: Menstruation* (Farrar, Straus and Giroux, 1999).

Stine Jensen, "De ernstige missie van de menstruatiemeisjes" [The serious mission of the menstruation girls], *NRC*, August 20, 2021, pressreader.com/netherlands/nrc-handelsblad/20210820/281784222172898.

Maarten Keulemans, "Misschien verstoren vaccins de menstruatiecyclus" [Vaccines may affect the menstrual cycle], de Volkskrant, September 6, 2021, volkskrant.nl/nieuws-achtergrond/misschien-verstoren-vaccins-de-menstruatiecyclus-kan-dat-kwaad~b885a8eb.

Nadia Kounang, "What's in your pad or tampon?," CNN, November 13, 2015, cnn.com/2015/11/13/health/whats-in-your-pad-or-tampon/index.html.

Vojka Lebar et al., "The effect of COVID-19 on the menstrual cycle: A systematic review," Journal of Clinical Medicine 11, no. 13 (July 2022): 3800, ncbi.nlm.nih.gov/pmc/articles/PMC9267255.

V. Male, "Menstrual changes after COVID-19 vaccination," BMJ 374, no. 2211 (2021), bmj.com/content/374/bmj.n2211.

Office on Women's Health, "Polycystic ovary syndrome," US Department of Health and Human Services, February 22, 2021, womenshealth.gov/a-z-topics/polycystic-ovary-syndrome.

Nadya Okamoto, Period Power: A Manifesto for the Menstrual Movement (Simon & Schuster Books for Young Readers, 2018).

"Painful periods (dysmenorrhea)," UC San Diego, studenthealth.ucsd.edu/resources/health-topics/painful-periods/index.html.

Jerilynn C. Prior, "Very heavy menstrual flow," Centre for Menstrual Cycle and Ovulation Research, October 4, 2017, cemcor.ubc.ca/resources/very-heavy-menstrual-flow.

Aishwarva Rohatgi and Sambit Dash, "Period poverty and mental health of menstruators during COVID-19 pandemic: Lessons and implications for the future," Frontiers in Global Women's Health 4 (2023): 1128169, ncbi.nlm.nih.gov/pmc/articles/PMC10014781.

"Sex and politics: Less taboo than periods?," Plan International, plancanada.ca/stories/letstalkperiods-on-menstrual-hygiene-day.

Honorata van den Akker and Lieke Smets, *Ben je ongesteld of zo? Alles wat je moet weten (en meer) over je menstruatie #nofilter* [Are you on your period or something? Everything you want to know (and more) about your period] (Ambo Anthos, 2021).

Honorata van den Akker and Lieke Smets, *Menstruatiemeisjes* [Menstruation girls], podcast, 31 episodes between October 29, 2020, and December 24, 2021.

Lukas van der Storm, "'Wrang als vrouwen moeten kiezen tussen groenten of tampons': Manifest eist actie van gemeente tegen menstruatiearmoede" ["It's ironic when women have to choose between vegetables and tampons": Manifesto demands action from the municipality against period poverty], *Trouw*, September 14, 2021, trouw.nl/economie/wrang-als-vrouwen-moeten-kiezen-tussen-groenten-of-tampons-manifest-eist-actie-van-gemeente-tegen-menstruatiearmoede~b391ac72.

Corien van Zweden, *Borsten: De levensloop van een intiem lichaamsdeel* [Breasts: The biography of an intimate body part] (De Bezige Bij, 2019).

Alex Whiting, "Women use 5,000 euphemisms to ease pain of talking about periods," *Reuters*, March 1, 2016, reuters.com/article/world/women-use-5-000-euphemisms-to-ease-pain-of-talking-about-periods-idUSL8N1693K7.

Chloe Williams, "Female homelessness and period poverty," National Organization for Women, January 22, 2021, now.org/blog/female-homelessness-and-period-poverty.

Chapter 3

"Anticonceptie," Rijksinstituut voor Volksgezondheid en Milieu, November 1, 2016, lci.rivm.nl/sites/default/files/bestanden/draaiboek-consult-seksuele-gezondheid/11-anticonceptie.pdf.

Mirjam J. A. Apperloo, Patti Vink, and Sjors Oosterbaan-Schram, "Zwangerschapsafbreking voordat de zwangerschap echoscopisch zichtbaar is: Wenselijk, haalbaar en veilig in

Nederland?" [Termination of pregnancy before the pregnancy can be seen by ultrasound examination: Desirable, feasible and safe in the Netherlands?], *Nederlands Tijdschrift voor Geneeskunde* 165, no. 43 (October 29, 2021), ntvg.nl/artikelen/zwangerschapsafbreking-voordat-de-zwangerschap-echoscopisch-zichtbaar.

Lynn Berger, "De donkere geschiedenis van de pil die de vrouw bevrijdde" [The dark history of the pill that liberated women], *De Correspondent*, October 28, 2014, decorrespondent.nl/1966/de-donkere-geschiedenis-van-de-pil-die-de-vrouw-bevrijdde/cd55de44-de3c-02de-2d6a-5e283f659c60.

"Birth control failure rates: The Pearl Index explained," *Drugs.com*, December 7, 2023, drugs.com/medical-answers/birth-control-failure-rates-pearl-index-explained-3554953.

Boston Women's Health Book Collective, *Our Bodies, Ourselves: A Book by and for Women* (Simon & Schuster, 1973).

Quirine Brouwer, "Steeds meer vrouwen stoppen met de pil of spiraaltje: '12 jaar lang niet mezelf geweest'" [Increasing numbers of women are stopping using the pill or coil: "I wasn't myself for 12 years"], *Metro*, September 26, 2021, metronieuws.nl/lifestyle/health-mind/2021/09/stoppen-met-pil.

Sonja Damstra-Wijmenga, *In smart zult gij uw kinderen baren: Opmerkelijke opvattingen over voortplanting* [In sorrow thou shalt bring forth children: Remarkable views on reproduction] (Boom/Belvédère, 1995).

"De anticonceptiekeuzehulp" [Tips on choosing contraceptives], *Seksualiteit.nl*, seksualiteit.nl/anticonceptietool.

Jonathan Eig, *The Birth of the Pill: How Four Crusaders Reinvented Sex and Launched a Revolution* (Pan Macmillan, 2014).

"Fewer women on the pill, IUDs more popular," Statistics Netherlands, June 19, 2014, cbs.nl/en-gb/news/2014/25/fewer-women-on-the-pill-iuds-more-popular.

"Key statistics for ovarian cancer," American Cancer Society, accessed July 2024, cancer.org/cancer/types/ovarian-cancer/about/key-statistics.html.

Kate Muir, "Why are women turning to TikTok for advice about the pill? Because doctors won't listen to us," *Guardian*, September 11, 2023, theguardian.com/commentisfree/2023/sep/11/women-tiktok-pill-doctors-social-media-contraception.

"Ovarian cancer statistics," Cancer Research UK, accessed July 2024, cancerresearchuk.org/health-professional/cancer-statistics/statistics-by-cancer-type/ovarian-cancer.

Emma Reilly, "Vasectomy 'a pretty easy' choice to make, says Hamilton dad," *Hamilton Spectator*, June 28, 2019, thespec.com/life/vasectomy-a-pretty-easy-choice-to-make-says-hamilton-dad/article_a64da871-a305-5a8a-9e57-4ceb73cda81f.html.

Stichting Anticonceptie Nederland [Dutch Foundation for Contraception], anticonceptie-online.nl.

"The world's abortion laws," Center for Reproductive Rights, accessed July 2024, reproductiverights.org/maps/worlds-abortion-laws.

Chapter 4

"Cancer stat facts: Uterine cancer," National Cancer Institute, accessed July 2024, seer.cancer.gov/statfacts/html/corp.html.

"Cervical cancer," Public Health Agency of Canada, May 2, 2024, canada.ca/en/public-health/services/chronic-diseases/cancer/cervical-cancer.html.

"Cervical cancer statistics," Cancer Research UK, accessed July 2024, cancerresearchuk.org/health-professional/cancer-statistics/statistics-by-cancer-type/cervical-cancer.

"Cervical cancer statistics," US Centers for Disease Control and Prevention, accessed July 2024, cdc.gov/cervical-cancer/statistics/index.html.

Esther H. Chen et al., "Gender disparity in analgesic treatment of emergency department patients with acute abdominal pain," *Academic Emergency Medicine* 15, no. 5 (May 2008): 414–18, pubmed.ncbi.nlm.nih.gov/18439195.

Elinor Clegborn, "The gender pain gap has gone on for too long—it's time we closed it," *New Scientist,* June 23, 2021, newscientist.com/article/mg25033400-100-the-gender-pain-gap-has-gone-on-for-too-long-its-time-we-closed-it.

De Gynaecoloog [The gynecologist], degynaecoloog.nl.

Hannah Devlin, "Women in UK waiting almost nine years for endometriosis diagnosis, study finds," *Guardian*, March 4, 2024, theguardian.com/society/2024/mar/04/women-in-uk-waiting-almost-nine-years-for-endometriosis-diagnosis-study-finds.

Lena Dunham, "In her own words: Lena Dunham on her decision to have a hysterectomy at 31," *Vogue*, February 14, 2018, vogue.com/article/lena-dunham-hysterectomy-vogue-march-2018-issue.

Endometriose Stichting [Endometriosis Foundation], endometriose.nl.

"Endometriosis," World Health Organization, March 24, 2023, who.int/news-room/fact-sheets/detail/endometriosis.

Mirjam Kaijer, *Ik ben geen man! Waarom zijn er zoveel vrouwen met onverklaarde gezondheidsklachten?* [I'm not a man! Why are there so many women with unexplained health issues?] (Lucht, 2021).

Cobie Lutters, "Laten we dit taboe doorbreken... We zijn het waard dat erover gesproken wordt" [Let's break this taboo... We need to talk about it], *Buikspreker: Endometriose Magazine*, August 2021.

Henk Maassen, "Komt een man of vrouw bij de dokter" [When a man or woman goes to the doctor], *Medisch Contact*, November 11, 2021, medischcontact.nl/actueel/laatste-nieuws/artikel/komt-een-man-of-een-vrouw-bij-de-dokter....

Hilary Mantel, *Giving Up the Ghost* (Fourth Estate, 2010).

Wim Schepens, Marieke Dermul, and Guy De Troyer, "Ellen Andries maakt een film over haar ziekte endometriose: 'Ik voelde mij dagelijks rotslecht, fysiek en mentaal,'" [Ellen Andries produced a film about her endometriosis: "I felt absolutely awful, both physically and mentally, every single day"], *Vrt.be*, June 9, 2020, vrt.be/vrtnws/nl/2020/07/08/documentairemaakster-over-haar-ziekte-endometriose.

"Uterine cancer statistics," Cancer Research UK, accessed July 2024, cancerresearchuk.org/health-professional/cancer-statistics/statistics-by-cancer-type/uterine-cancer.

Inez van Dullemen, *Een schip vol meloenen* [A ship full of melons] (De Bezige Bij, 2017).

Chapters 5 and 6

"Archived—Changing fertility patterns: Trends and implications," Health Canada, 2005, canada.ca/en/health-canada/services/science-research/reports-publications/health-policy-research/changing-fertility-patterns-trends-implications.html.

"Assisted vaginal birth (ventouse or forceps)," Royal College of Obstetricians and Gynaecologists, April 2020, rcog.org.uk/for-the-public/browse-our-patient-information/assisted-vaginal-birth-ventouse-or-forceps.

E. P. Backes and S. C. Scrimshaw, eds., *Birth Settings in America: Outcomes, Quality, Access, and Choice* (National Academies Press, 2020), ncbi.nlm.nih.gov/books/NBK555484.

R. W. Bakker and A. T. M. Verhoeven, *Geschiedenis van de verloskunde en gynaecologie in Nederland* [A history of obstetrics and gynaecology in the Netherlands] (NVOG, 2013).

Floor Bal, "Arts versus vroedvrouw" [Doctor versus midwife], *Historisch Nieuwsblad*, April 27, 2010, historischnieuwsblad.nl/arts-versus-vroedvrouw.

Joshua H. Barash, "Diagnosis and management of ectopic pregnancy," *American Family Physician*, July 1, 2014, aafp.org/pubs/afp/issues/2014/0701/p34.html.

Ana Pilar Beltrán et al., "The increasing trend in caesarean section rates: Global, regional and national estimates: 1990–2014," *PloS One* 11, no. 2 (2016): e0149343, ncbi.nlm.nih.gov/pmc/articles/PMC4743929.

Mireia Bernal Claverol et al., "Maternal, perinatal and neonatal outcomes of triplet pregnancies according to chorionicity: A systematic review of the literature and meta-analysis," *Journal of Clinical Medicine* 11, no. 7 (April 2022): 1871, ncbi.nlm.nih.gov/pmc/articles/PMC8999732.

"Births in England and Wales by characteristics of birth 2: 2013," Office for National Statistics, November 17, 2014, ons.gov.uk/peoplepopulationandcommunity/birthsdeathsandmarriages/livebirths/bulletins/characteristicsofbirth2/2014-11-17.

"Canada's fertility rate reaches an all-time low in 2022," Statistics Canada, January 31, 2024, www150.statcan.gc.ca/n1/daily-quotidien/240131/dq240131c-eng.htm.

Sonja Damstra-Wijmenga, *In smart zult gij uw kinderen baren: Opmerkelijke opvattingen over voortplanting* [In sorrow thou shalt bring forth children: Remarkable views on reproduction] (Boom/ Belvédère, 1995).

Arie de Graaf, "Vruchtbaarheid in de twintigste eeuw" [Fertility in the twentieth century], Centraal Bureau voor de Statistiek, March 26, 2008, cbs.nl/nl-nl/achtergrond/2008/13/vruchtbaarheid-in-de-twintigste-eeuw.

Nadia Filippini, *Pregnancy, Delivery, Childbirth: A Gender and Cultural History From Antiquity to the Test Tube in Europe* (Routledge, 2020).

"Geen moeder, maar ouder: Zwangere non-binaire transpersoon stapt naar rechter" [Not mother, but parent: Pregnant nonbinary trans person goes to court], *NOS*, August 19, 2021, nos.nl/ artikel/2394398-geen-moeder-maar-ouder-zwangere-non-binaire-transpersoon-stapt-naar-rechter.

Robyn Horsager-Boehrer, "Pregnancy complications: What are the chances that they'll happen again?," UT Southwestern Medical Center, January 29, 2019, utswmed.org/medblog/ recurrent-pregnancy-complications/.

Donna L. Hoyert, "Maternal mortality rates in the United States, 2022," National Center for Health Statistics, May 2024, cdc. gov/nchs/data/hestat/maternal-mortality/2022/maternal-mortality-rates-2022.pdf.

Randi Hutter Epstein, *Get Me Out: A History of Childbirth From the Garden of Eden to the Sperm Bank* (W. W. Norton & Company, 2010).

Nikit Kadam et al., "Odds and predictors of monozygotic twinning in a multicentre cohort of 25,794 IVF cycles," *Journal of Clinical Medicine* 12, no. 7 (April 2023): 2593, ncbi.nlm.nih.gov/ pmc/articles/PMC10095500.

Julie Y. Lo, "What moms should know about forceps and vacuum deliveries," UT Southwestern Medical Center, April 19, 2022, utswmed.org/medblog/forceps-vacuum-delivery.

"Low-risk cesarean sections," Canadian Institute for Health Information, November 2023, cihi.ca/en/indicators/low-risk-caesarean-sections.

"Maternal death rates in the UK have increased to levels not seen for almost 20 years," University of Oxford, January 11, 2024, ox.ac.uk/news/2024-01-11-maternal-death-rates-uk-have-increased-levels-not-seen-almost-20-years.

Karen Miles, "What to know if your baby is breech," *Babycenter*, November 1, 2021, babycenter.com/pregnancy/your-body/breech-birth_158.

Keith Moore, "How accurate are 'due dates'?," *BBC News*, February 3, 2015, bbc.com/news/magazine-31046144.

Franz Karl Naegele, *Leerboek der verloskunde voor vroedvrouwen* [A textbook of obstetrics for midwives] (Noordendorp, 1837), books.google.ca/books/about/Leerboek_der_verloskunde_voor_vroedvrouw.html?id=iWbpPlLzjzgC&redir_esc=y.

R. Orvieto et al., "Does mRNA SARS-CoV-2 vaccine influence patients' performance during IVF-ET cycle?," *Reproductive Biology and Endocrinology* 19, no. 1 (May 2021), pubmed.ncbi.nlm.nih.gov/33985514.

"Overview of multiple pregnancy," Stanford Medicine, accessed July 2024, stanfordchildrens.org/en/topic/default?id=overview-of-multiple-pregnancy-85-P08019.

Justina Petrullo, "US has highest infant, maternal mortality rates despite the most health care spending," *AJMC*, January 31, 2023, ajmc.com/view/us-has-highest-infant-maternal-mortality-rates-despite-the-most-health-care-spending.

"Premature labour and birth," NHS, January 10, 2024, nhs.uk/pregnancy/labour-and-birth/signs-of-labour/premature-labour-and-birth.

Gopal K. Singh, "Maternal mortality in the United States, 1935–2007," US Department of Health and Human Services, October 2010, researchgate.net/publication/335432496.

Beatrijs Smulders, *Bloed: Een vrouwengeschiedenis* [Blood: A herstory] (Nijgh & Van Ditmar, 2021).

Beatrijs Smulders and Mariël Croon, *Veilig bevallen: Het complete handboek* [Safe delivery: The comprehensive guide] (Kosmos Z&K, 1996).

Eric A. P. Steegers, ed., *Textbook of Obstetrics and Gynaecology: A Life Course Approach* (Bohn Stafleu van Loghum, 2019).

Sanne I. Stegwee et al., "Prognostic model on niche development after a first caesarean section: Development and internal validation," *European Journal of Obstetrics & Gynecology and Reproductive Biology* 283 (2023): 59–67, ejog.org/article/S0301-2115(23)00022-2/pdf.

Joan Stephenson, "Rate of first-time cesarean deliveries on the rise in the US," *JAMA Network*, July 12, 2022, jamanetwork.com/journals/jama-health-forum/fullarticle/2794350.

"Twins and more," NHS University Hospitals Sussex, February 7, 2024, uhsussex.nhs.uk/services/maternity/pregnancy/multiples.

Rineke van Daalen, "De groei van de ziekenhuisbevalling: Nederland en het buitenland" [The growth of hospital birth: The Netherlands and abroad], *Amsterdams Sociologisch Tijdschrift* 15, no. 3 (December 1988): 414–45.

Sander Voormolen, "Beginnen in een kunstbaarmoeder" [Starting in an artificial womb], *NRC*, September 28–29, 2021.

Steven S. Witkin and Larry J. Forney, eds., "The microbiome and women's health," themed issue, *BJOG: An International Journal of Obstetrics and Gynaecology* 127, no. 2 (December 2019): doi. org/10.1111/1471-0528.16010.

Robert Witte, "4 transmannen over zwangerschap" [4 trans men on pregnancy], *Trans,* transmagazine.nl/transman-en-zwangerschap.

Chapter 7

"But I am only 40! Can this be menopause?," *Menopause and U,* accessed July 2024, menopauseandu.ca/am-i-in-menopause/40-can-menopause.

Burcu Cevlan and Nebahat Özerdoğan, "Factors affecting age of onset of menopause and determination of quality of life in menopause," *Turkish Journal of Obstetrics and Gynecology* 12, no. 1 (March 2015): 43–49, ncbi.nlm.nih.gov/pmc/articles/PMC5558404.

Elles de Bruin, *Opvliegers* [Hot Flashes], podcast, NPO Radio 1 and Omroep Max, seven episodes between October 21, 2020, and December 7, 2020, nporadio1.nl/podcasts/opvliegers.

Herman Depypere and Sofie Vanherpe, *Menopauze: Alle vragen beantwoord* [Menopause: All questions answered] (Borgerhoff & Lamberigts, 2019).

Isa Hoes and Medina Schuurman, *Te lijf: De kunst van het mooi ouder worden* [The art of ageing gracefully] (Ambo Anthos, 2016).

Jonathan Lambert, "Living near your grandmother has evolutionary benefits," *Goats and Soda* (NPR), February 7, 2019, npr.org/sections/goatsandsoda/2019/02/07/692088371/living-near-your-grandmother-has-evolutionary-benefits.

Richard A. Morton, Jonathan R. Stone, and Rama S. Singh, "Mate choice and the origin of menopause," *PLoS Computational Biology* 9, no. 6 (June 13, 2013), pubmed.ncbi.nlm.nih.gov/23785268.

Louise R. Newson, "Best practice for HRT: Unpicking the evidence," *British Journal of General Practice* 66 (December 2016): 597–98, ncbi.nlm.nih.gov/pmc/articles/PMC5198659.

José Rozenbroek and Jos Teunis, *De overgang: Het no-nonsense handboek* [Menopause: The no-nonsense guide] (Atlas Contact, 2018).

Neda Sarafrazi et al., "Osteoporosis or low bone mass in older adults: United States, 2017–2018," National Center for Health Statistics, Data Brief no. 405 (March 2021), cdc.gov/nchs/products/databriefs/db405.htm.

"Symptoms and signs of miscarriage," American Pregnancy Association, accessed July 2024, americanpregnancy.org/getting-pregnant/pregnancy-loss/signs-of-miscarriage.

Megan Tatum, "Without support, many menopausal workers are quitting their jobs," BBC, April 9, 2024, bbc.com/worklife/article/20240408-menopause-women-job-quits.

Marieke van Twillert, " 'De overgang is een echte carrièrekiller' " ["Menopause is a real career killer"], *Medisch Contact*, October 20, 2021, medischcontact.nl/actueel/laatste-nieuws/artikel/de-overgang-is-een-echte-carrierekiller.

Janneke Wittekoek and Dorenda van Dijken, *Hart and Hormonen: Fit de overgang in* [Heart and hormones: Fit through menopause] (Lucht, 2020).

Index

Illustrations indicated by page numbers in italics